The Grammar of Genes

European Semiotics: *Language, Cognition, and Culture*
Sémiotiques Européennes: *langage, cognition et culture*

Edited by / Série rédigée par
Per Aage Brandt (Aarhus), Wolfgang Wildgen (Bremen/Brême),
and/et Barend van Heusden (Groningen/Groningue)

Volume 6

PETER LANG
Bern · Berlin · Bruxelles · Frankfurt am Main · New York · Wien

Ángel López-García

The Grammar of Genes

How the Genetic Code Resembles
the Linguistic Code

PETER LANG
Bern · Berlin · Bruxelles · Frankfurt am Main · New York · Wien

Bibliographic information published by Die Deutsche Bibliothek
Die Deutsche Bibliothek lists this publication in the Deutsche Nationalbibliografie;
detailed bibliographic data is available on the Internet at ‹http://dnb.ddb.de›.

British Library and Library of Congress Cataloguing-in-Publication Data:
A catalogue record for this book is available from *The British Library*, Great Britain,
and from *The Library of Congress*, USA

ISSN 1423-5587
ISBN 3-03910-654-6
US-ISBN 0-8204-7171-2

© Peter Lang AG, European Academic Publishers, Bern 2005
Hochfeldstrasse 32, Postfach 746, CH-3000 Bern 9
info@peterlang.com, www.peterlang.com, www.peterlang.net

All rights reserved.
All parts of this publication are protected by copyright.
Any utilisation outside the strict limits of the copyright law,
without the permission of the publisher, is forbidden and liable to prosecution.
This applies in particular to reproductions, translations, microfilming,
and storage and processing in electronic retrieval systems.

Index

Acknowledgements 9

1 *Dichotomic thought* 11

2 *Inside out: the evolution of the brain* 15
   2.1 The battle over linguistic evolution:
         the Darwinian paradigm 15
   2.2 Baldwinian evolution and exaptation 18
   2.3 The big mutation hypothesis 21
   2.4 In search of another explanation 24

3 *Outside in: the evolution of language* 27
   3.1 Learning a language 27
   3.2 Two outside in models: motor control and gesture 28
   3.3 The evolution of language culture 30
   3.4 An old fashioned comparison: language as an organism 31
   3.5 A new comparative approach: language as a species 34
   3.6 The formal properties of linguistic evolution 35
         3.6.1 The structure of evolution 35
         3.6.2 Methodology: samples for the study of evolution 36
         3.6.3 The causes of evolution 36
   3.7 Linguistic speciation 38
         3.7.1 Mechanisms of speciation 38
         3.7.2 Patterns and rythms of speciation 39
   3.8 Language is culture but grammatical units
         are not cultural ones 39

4 *Formal inheritance* 43
   4.1 Do formal patterns precede their manifestation? 43
   4.2 The inheritance of form in the natural world 46
   4.3 The genes of language? 48
   4.4 On pregnance and salience: the concept
         of 'figurative effect' 49

    4.5 Linguistic pregnances: the emergence of syntactic laws 51
    4.6 Epigenesis as a source for emergent properties 56
    4.7 Protolanguage as a product of perceptual networks 58

5 *How complex syntax can be?* 63
    5.1 On linguistic form 63
    5.2 Types of grammar: generative, cognitive,
         and functional grammars 65
    5.3 The formal syntax 68
    5.4 A glance at the syntactic principles 69
         5.4.1 Dependence relations characterize
             the summing up of lexical items 69
         5.4.2 Dependence relations are established
             among heterogeneous units 70
         5.4.3 Agreement: some overt relationships
             help to solve adjustment problems 71
         5.4.4 Anaphora and ellipsis as cohesive procedures 72
         5.4.5 Categories 74
         5.4.6 Movement relates different orderings
             of the chain of words 74
         5.4.7 Restrictions of movement 76
         5.4.8 The construction of texts in speech acts forms 76
    5.5 On the significance of the formal properties of syntax 77

6 *A blind alley* 81
    6.1 A striking parallelism 81
    6.2 Some inadequacies of the comparison 84
    6.3 What does the genetic code really look like
         and how does it work? 89

7 *On the code: the form of genetic code maps
   the form of linguistic code* 95
    7.1 Method: looking for a formal correspondence 95
    7.2 Phrase structure and Codon structure 97
         7.2.1 The third base and the Complement 97
         7.2.2 The second base and the Head 98
         7.2.3 The first base and the Specifier 99
         7.2.4 Summary: a codon is formally similar to a Phrase ... 99
    7.3 Nucleotides and categories 101

8 *Further formal parallelisms between genetic code and linguistic code* 105
   8.1 A provisiuonal summary of formal resemblances 105
   8.2 Satellite DNA as a formal pattern for constituent structure ... 106
   8.3 Genetic crossing-over resembles syntactic movement 109
   8.4 Crossover fixation as a formal model of recursion 110
   8.5 Wobble as a formal schema for agreement 111
   8.6 Empty categories follow the formal model of transposons 115
   8.7 Subjaceny is framed like cis-dominance 118

9 *Linguistic texts and genomic strings share some formal devices* 121
   9.1 On genetic levels and linguistic levels 121
   9.2 Sentences as transcriptional units 124
        9.2.1 The promoter and the talk openings 125
        9.2.2 Control at termination and the talk closings 127
   9.3 The utterance markers and the translation markers 129
   9.4 The regulation of texts and the operon 131
   9.5 Positive and negative control 135
        9.5.1 Positive control 136
        9.5.2 Negative control 137

10 *The organizer* 141
   10.1 The inheritance of formal codes 141
   10.2 The limits of emergence 144
   10.3 The concept of pre-program 147
   10.4 The organizer 149
   10.5 Homeobox once again 152
   10.6 Genetic code meets linguistic code 153

11 *The limits of complexity* 157
   11.1 A double source for complexity 157
   11.2 The origin of language: a gradual process with a break 160
   11.3 Syntax as a symbiosis of several genetic encodages 164

12 *Some concluding remarks* 169

References 175

Index 181

Acknowledgements

This research was supported by grant BFF2003-05981 from the Ministerio de Ciencia y Tecnología of Spain. It includes many of the issues I first raised in a Spanish book, *Fundamentos genéticos del lenguaje*: I thank the editor, Mrs. Iosune García, for her support. I am also grateful to some academic meetings I had occasion to benefit from: I had the opportunity to present an early draft of Chapter 3 at the conference on Linguistic Variation that was organized by Jens Luedtke at the University of Heidelberg in November 2002, and a preliminary version of chapter 4 at the conference on Actantiality that was organized by Gerd Wotjak at the University of Leipzig in October 2003. I was also able to discuss many aspects of Chapters 6–9 with the students of my 2003–2004 course on Current Trends in Linguistics at the University of Valencia. I benefited a lot from discussions with the colleagues that joined a German presentation of the main ideas in a conference I held at the University of Mainz in June, 2004. I joined the biomedical symposium on Biology and Language that was organized by Martin Caicoya in Valdediós (Asturias, Spain) in September, 2004: it gave me the opportunity to contrast my ideas with the other side of the mirror. I am grateful to Mr. Desmond Donnellan and Mrs. Fabiola Barraclough who made corrections in English grammar and style. Finally, thanks are also due to Per Aage Brandt and Wolfgang Wildgen who helped get this project off the ground and the editorial support they gave to it.

1 Dichotomic thought

Current approaches to the problem of the origin of language face up to its dichotomous character. It seems that in relation to this issue people are forced to choose between two alternatives. The table sums up the most remarkable dichotomies:

language from inside	*language from outside*
cognitive	communicative
individual	social
innate	learned
biological	cultural
structural	functional
modular	general

Researchers that consider language to be a phenomenon that evolved essentially within the human brain generally emphasize its *cognitive* status: hence, language would be a hard-wired specific (i.e. *modular*) capacity which developed in some *individual* hominids, and which under standard *biological* processes of natural selection imprinted an *innate* capacity to acquire linguistic *structures* in the genome of their descendants. Thus, a very comfortable and rather unproblematic picture results when we accept an "inside out explanation" for the origin of language.

On the contrary, the outside in explanation also looks quite convincing. Let us suppose language was a *communicative* ability developed by a group of hominids to increase their *social* cohesion and facilitate certain survival *functions* such as hunting, tool making, or the problem of feeding infants: consequently, language was related to these *cultural* meanings from the beginning, and the descendants of the hominid group had to *learn* it every time they learned to hunt or to make instruments using their enlarged *general* intelligence.

Such embarrassing situations are very common in science. Frequently two paradigms aim to explain a particular issue by adopting opposite explanations, both seem capable of justifying most of the observed empirical facts. Ptolemy supported the view that the earth was the centre of the solar system, while Kepler constituted the sun as its central point, albeit both sets of formal statements enabled their respective practitioners to satisfactorily explain the behavior of the planets. Nevertheless, after an exhaustive confrontation between the solutions of both competing proposals, we find that the first failed to explain some facts of nature, and was thus rejected as invalid. Similarly, the flogist theory of burning or the taxonomical explanation of the differences among living species were definitively abandoned and belong to the history of science.

But it is equally possible that two opposing theories both prove to be right, despite the fact that each of them adopts a solution contrary to the other. This was the case with the corpuscular and ondulatory theories of light: the first assumes that light is made out of particles, the second, that it is composed of waves; we now know that it is both. How do we manage such a situation?: it would make no sense to reject either one of the competing proposals, on the contrary we are forced to accept both by finding a picture of the world within which they are compatible. This seems to be the case for both the inside and outside theory on the origin of language.

There are some empirical data that suggest a conciliation between the left and the right positions in the table above. For example, Aiello & Dunbar (1993) demonstrated that there is a relationship between neocortex size and social group size, not only among primates and other animals, but also among humans: this means that cognitive and communicative abilities of hominids developed at the same time, and that there was a close relationship between them. One could object and say that the size of the brain is related to general intelligence, and that this correlation supports only the communicative basis of language origin (in fact, Dunbar thinks that the principal function of language is to enable the exchange of social information – gossip –). However, as alleged by Ulbaek (1998), the expansion of communicative skills is strictly narrowed because a selfish animal has very few things to communicate, whereas cognition is the other way around for extracting information from the environment is inherently selfish: this implies a central role for cognition in the development of language, although we needn't support Ulbaek's opinion that it evolved from animal cognition and not from animal communication.

Another bridging approach links the opposite terms social/individual and functional/structural together. Notice that despite vision or manual ability, which belong to the individual domain, language does not exist outside the social group. Humans see the world and handle the objects of the world alone, but they cannot speak of the world alone, they always relate something to other humans about the world. Hence language could not be born as the result of a genetic change in a single individual otherwise it would lack any adaptive fitness at all. Language was born to facilitate the social information exchange – whether cognitive, or cohesive, or both – within the group. This means that the emergence of language presupposes the previous existence of some social patterns of a mimic (Donald, 1991) or of a motor gestural nature (Armstrong, Stokoe & Wilcox, 1995) among hominids, which helped them to understand another individual's mental state. However, adaptation only began when the best patterns were selected out of a set, and transmited to the descendants: this process is necessarily an individual one. And whatsmore: although the emergence of a given pattern was originally due to the fact that it fulfilled a cognitive or communicative function (Givón, 2001) when it was interiorized by the individual as a hardwired web of neural connections, it converted into a formal pattern, that is, into a structure.

Neither do the remaining options seem insurmontable. Is language an innate capacity or is it a learned one? The state of art usually opposes the innatist position by Chomsky, and the generative grammarians, to the behaviorist hypothesis that is held by the remaining linguistic schools, and also by the great majority of psychologists or anthropologists. But, for me, the question seems wrongly posed this way. Everyone is forced to recognize that humans learn languages from other humans (be it their parents, their teachers, or the dustman) but, at the same time, that they are able to do so precisely because they are humans. In other words, language is both, learned and innate. The problem is not what language *is*, but *how* language emerged on earth. If it was culturally, then language developed as a semiotic system in the social group and it finally became incorporated into the human genome as a specific ability that other animals do not share. If it was biologically, then language had to manifest as a multipurpose ability in the cultural domain, which converted it into a vehicle for the expression of cultural meaning. Thus, the modular vs. general character of human linguistic intelligence seems to be a rather illusive question: social meanings (which obviously exclude the inner thoughts and the emotions) do not exist before they are linguistically expressed, and language, according to the aforism by Wittgenstein – *Der Satz zeigt die logische Form der Wirklichkeit* –, goes beyond any meaning. In this sense, language probably originated (Mithen, 1996) as an intermediate structure that joined the social mind, the technological mind, and the natural world mind together. Language is, then, specific (i.e. modular) and overwhelming (i.e. general) at the same time.

The time has come to put the dichotomic thought on language origin away. With this in mind, there are not two irreducible positions, but rather two complementary perspectives: the *inside out* approach and the *outside in* approach. This does not mean, however, that either of them are without problems. If language originated inside the human brain, we have to explain just how a property emerges which, in turn, distinguishes humans from any other animal on earth. If, on the contrary, language developed first as a communicative device of the social group of hominids, we have to explain how it is embodied in form of a very different structure. The problem of the origin of language seems to be, then, of a biological nature in any case: from inside the brain, we have to explain how could a continuous Darwinian adaptation that produced a very different species, the *homo loquens*, took place; and from outside the brain, on the contrary, we have to justify how could a radical, discontinuous situation, the gap that separates culture from nature, could be bridged for humans who are, after all, just another species.

It is important to point out this biological aspect, which should never be left aside, because scholars that work from the outside in perspective are used to thinking that the origin of the language problem can be solved entirely by human sciences. This is not the case. If we adopt an inside out approach, then we will get more help from Genetics and from Physics; if, on the contrary, we adopt an outside in approach, then we will get more help from Psychology and from Semiotics. But *more* does not mean *complete*: no matter how language appeared on earth, it is a faculty that only humans possess, and, thereby, a capacity of our species. Certainly it only develops in the social group, but it exists – as Ferdinand de Saussure pointed out – in every individual. When the group vanishes, then their political, religious, or esthetical ideas disappear at the same time; language, however, remains as long as the people that speak it are alive.

2 Inside out: the evolution of the brain

2.1 The battle over linguistic evolution: the Darwinian paradigm

The publication of Charles Darwin's essay on the origin of species in 1859 turned a lot of well established ideas in Biology upside down. For instance, Darwin demonstrated not only the evidence of evolution, but also that there is a natural cause for evolution as such: natural selection. According to Darwin, changes that are experienced by living beings have some properties in common with the inanimate world: animals do not develop from their ancestors in the same way as stones develop from previous stones, but both processes are due to a set of natural forces that follow the principle of causality. It is known how seriously such a position challenged the theological concept of the divine watchmaker and how many difficulties with the theory of evolution had to be surmounted before being accepted as a respectable component of the paradigm of natural science.

At the present time there are no longer any political objections, but scientific queries relative to the theory of evolution still remain. Some of them belong to the current paradigm of Biology and will not be discussed here. We know, Darwin did not know what factor caused hereditary variation because of the common belief that parental features are mixed in every meiosis and, consequently, they would disappear in very few generations. This was obviously not the case, but the explanation was only available after Mendel postulated his famous laws. Mendel recognized a 'particulate factor', the gene, which passes unchanged from parent to progeny: according to this view, his first law establishes the segregation of alleles, and his second law the independent assortment of genes. Nevertheless, it seems that Darwin did not become aware of Mendel's theory and consequently did not consider that genetic factors are discrete.

Now, the discrete nature of genes is a well known fact in Biology, but meanwhile new problems have arisen: in order to explain evolution, the genes must be altered to some extent, making it possible for natural selection to slowly produce a modified organism. De Vries, the scholar that rediscovered Mendel's ideas, supposed that those changes were big ones, and called them mutations. However, as mathematical considerations by

Haldane (1993) and empirical evidence formerly adduced by Dobzhansky (1937, 1975) made clear, in the natural world there is no radical change at all, and the sudden appearance of a new species should be considered a 'hopeful monster story'. Be that as it may, the current theory of evolution claims that biologic change is always gradual, no matter whether the orthodox model of the synthesis extends gradually to any phase of the process in a regular manner, whereas the model of punctuated equilibrium, by Eldredge and Gould (1972), supposes that some periods of relatively fast changes, that may give rise to new species, are followed by rather long and essentially static periods of slow changes.

The Darwinian paradigm constitutes a solid endeavour of the human mind to understand the laws of nature. No wonder that, for the time being, it meets with natural scientists' approval. As R. Dawkins pointed out in his well known book *The selfish gene* (1976, 1):

> Intelligent life on a planet comes of age when it first works out the reason for its own existence. If superior creatures from space ever visit earth, the first question they will ask, in order to assess the level of our civilization, is: 'Have they discovered evolution yet?'. Living organisms have existed on earth, without ever knowing why, for over three thousand millions years before the truth finally dawned on one of them. His name was Charles Darwin.

But humans still get into trouble when trying to understand themselves. Darwin was aware of this fact, and devoted a second book, *The Descent of Man, and Selection in Relation to Sex* (1871), to demonstrate our descent from animals, and in particular apes. Modern discoveries in Paleontology and in Biology have proved him right. But there is still one problem left. Terrence W. Deacon (1997, 22–23) sets it out as follows:

> The question that ultimately motivates a perennial fascination with human origins is not who were our ancestors, or how they came to walk upright, or even how they discovered the use of stone tools. It is not really a question that has a paleontological answer. It is a question that might otherwise be asked of psychologists or neurologists or even philosophers. Where do human minds come from? The missing link that we hope to fill in by investigating human origins is not so much a gap in our family tree, but a gap that separates us from other species in general.

Is there really a gap? Sure, some species behave socially, most mammals experience fear and anger, some animals communicate with others, all of them perceive the external world [...], but it seems no animal possesses language. There is a gap, then, although Darwin was convinced otherwise.

He did not deal with the problem of the origin of language in his founding book of 1859, but the essay of 1871 was directly concerned with this. As we know, Darwin thought that language evolved from communicative attempts made by animals, such as shouts, gestures and cries, simply by acquiring more and more complex communicative skills with successive generations. Although now we know that this position was fundamentally wrong, the Darwinian paradigm still supports Biology as a science. There are certainly many animals that are able to communicate with each other. If you wish, you may consider that these communicative interactions follow the rules of a language. You must remember however, it is a very different language to ours, it is not symbolic, it encapsulates elements of reality by means of iconic signs. Human languages are symbolic, in contrast to animal 'languages' which are based on signs rather than symbols.

Thirty years ago scholars were optimistic about the potential progress to be made by teaching some symbolic systems like Ameslan (American Sign Language) to chimpanzees. In fact, experiments by Gardner and Gardner (1969), Premack (1971) and Rumbaugh (1977) succeeded in teaching one hundred 'words' and a few combinatory sequences to primates in the laboratory. However, these animals were very refractory in learning human speech any further than that. Current research on animal communication does not support the Darwinian approach and it has made clear that language is human specific. As Deacon (1997, 31–34) says:

> [...] although they are complex, these elaborate repertoires of calls, displays, and gestures do not seem to map onto any of the elements that compose languages. Although various researchers have suggested that parallels to certain facets of language are to be found in the learned dialects of birdsong, the external reference evident in vervet monkey alarm calls or honeybee dances, and the socially transmitted sequences of sounds that make up humpback whale songs [...], these and many other examples like them only exhibit a superficial resemblance to language learning, word reference, or syntax, respectively [...] This lack of precedent makes language a problem for biologists. Evolutionary explanations are about biological continuity, so a lack of continuity limits the use of the comparative method in several important ways.

In his book Deacon rejects so called hopeful monster explanations like the innatist proposal by Noam Chomsky, all of which are based on a gap. Curiously enough, however, he coincides with Chomsky (1996, 1–2) when emphasizing the difficulty that language represents for Biology:

> How can a system such as human language arise in the mind/brain, or for that matter, in the organic world, in which one seems not to find anything like the basic properties

of human language? That problem has sometimes been posed as a crisis for the cognitive sciences. The concerns are appropriate, but their locus is misplaced: they are primarily a problem for biology and the brain sciences, which, as currently understood, do not provide any basis for what appear to be fairly well established conclusions about language.

Nevertheless, Deacon does not give up, and tries to investigate the co-evolutionary exchange between language and brain over two million years. The way he does it is based on a former proposal that contributed to softening the paradigm of the synthesis: Baldwinian evolution.

2.2 Baldwinian evolution and exaptation

It is obvious that the emergence of language would be easier to understand if the learned abilities could be genetically transmitted to the next generation. Suppose that a group of hominids, e.g. some branch of Australopithecus, invent a rudimentary code for social interaction consisting of a limited set of shouts, each with its specific referential meaning (for feeding, sexual interchange, flea cleaning and so on), and a few combinatory rules that join them together. This minimum grammar could be called a protolanguage: in fact, as Derek Bickerton (1990) convincingly demonstrated a few years ago, this is the case with the communicative skills developed by chimpanzees trained to learn Ameslan; and, most interestingly, they do not differ from the skills that characterize the speech of children before the age of two. But such a grammar will not contribute to the evolution of protolanguage unless inherited by the next generation, because if the infant chimpanzee had to begin with a zero level of communicative skills again, no progress would be conceivable.

For evolution to be possible they had to start from the level of linguistic knowledge attained by their parents, modifying and improving it during the time of their own generation, and so on. Unfortunately, the inheritance of acquired features represents the foundation of an alternative hypothesis formulated by Lamarck, a contemporary scholar of Darwin, which was demonstrated to be false: our parents could have learned to play piano, to drive a car, or to dance tango, but these abilities do not benefit us anyway, as we need to learn them by ourselves from the beginning. This would also be the case with language: every chimpanzee generation develops some communicative skills, but none of them benefit from the communicative findings of their ancestors. A conclusion may be drawn: protolanguages cannot evolve in chimpanzees. And in fact they did not. The question is why the branch of Australopithecus that converted step by step in our own species, the Homo sapiens, apparently did so.

The answer of Deacon is based on Baldwinian evolution, a slight, but decisive, modification of Darwin's theory that was outlined by Mark Baldwin. Deacon (1998, 322) explains it as follows:

> Baldwin suggested that learning and behavioral flexibility can play a role in amplifying and biasing natural selection because these abilities enable individuals to modify the context of natural selection that affects the future kin. Behavioral flexibility enables organisms to move into niches that differ from those that ancestors occupied, with the consequence that succeeding generations will face a new set of selection pressures. For example, an ability to utilize resources from colder environments may initially be facilitated by seasonal migratory patterns, but if adaptation to this new niche becomes increasingly important, it will favor the preservation of any traits in subsequent generations that increase tolerance to cold, such as the deposition of subcutaneous fat, the growth of insulating hair, or the ability to hibernate during part of the year.

Baldwinian paradigm allows Darwin theory to incorporate some kind of learned features into genes: no wonder that it was accepted by Darwin himself when he observed that human children are born with a hardened soles which they obviously must have developed during the embryonic phase, and that they apparently inherited from those ancestors that acquired tough skin by walking on bare feet. The explanation of the emergence and evolution of language in the unprecedented work Terrence Deacon looks very alike. Suppose a group of primates has fortuitously developed a special ecological niche made out of new objects or processes they need to refer to. Their descendants grow up in a context that requires new symbolic abilities, but they do not find them by chance, as was the case with their parents, they have to learn them. Step by step the environment is becoming more and more complex and, at the same time, the primates brain is being forced by natural selection to accommodate such a situation. After many generations the interaction between mind and context will produce an enlarged brain that, especially through the neural connexions of the frontal lobe, is able to speak.

They were able to speak, but they did not. What contrasts humans with primates is that our ancestors, the hominids, did it. A proposal, that came to join Baldwinian evolution in an attempt to explain the origins of language yet without leaving the Darwinian paradigm, has recently been made by Calvin and Bickerton (2000): exaptation. Exaptation is the extension of the functional possibilities of an adaptive feature which evolved through natural selection. For example, feathers were first acquired by birds as an adaptation to heat retention, but now they are one of the most important components of their flight mechanism. Similarly, the bladder that permited fishes to float while swimming became the lungs which the birds that evolved from those fishes use for breathing and for flying.

The hypothesis of Calvin and Bickerton is very simple. Primates do not speak, but they have a complex social life that is well documented (Goodall, 1986). This social life includes internalised knowledge of social roles such as agent, patient or theme (I decide to protect you today, as an agent, because I remember you protected me, as a patient, yesterday, etc.). As social life became more and more complex, our ancestors incorporated this social calculus to the protolanguage they were using to communicate with each other. As a consequence, the protolanguage was converted in an adult language: after all, a sentence is a set of nouns that are semantically required by a verb together with the thematic structure that determines which kind of dependence they maintain in relation to it. For example, *John, Mary, look* (or, *Mary, John, look*, or *look, Mary, John*, any order of elements being possible) is a typical sentence of the protolanguage that apes may perform through gesture and which hominids probably performed vocally. However, *John*$_{Agent}$ *is looking at*$_{predicate}$ *Mary*$_{Patient}$ is a sentence of the English language because the thematic roles of the participants are manifested by means of some grammatical markers (by the order of elements, the agreement relationships, or the prepositions). These grammatical markers are very different in every language, but the abstract concepts they are related to remain essentially the same: for example, the subject-object relationship, that underlies the agent-patient relation with action verbs like *to look at*, is manifested in English by the order of words (*John looked at Mary* does not mean *Mary looked at John*), but in Spanish it is manifested by agreement morphemes and by the preposition *a* that precedes the object (*Juan miró a María* expresses the same situation as *a María miró Juan*, but both are different from *a Juan miró María*).

Exaptation lends a suitable background for many phenomena that took place when language was born, perhaps 100.000 years ago. Furthermore, the emergence of language presupposes the previous exaptation of other organic functions, for instance, the descent of the larynx, the lateral specialization of the brain, or its prefrontal expansion. Nevertheless, the exaptation of a social calculus, which certainly may have contributed to the birth of language, does not explain most of its formal properties, it only accounts for the thematic theory. Grammar continues to be a problem for evolutionists: on the one hand, the structure of the phrase, the rules that govern anaphora, the agreement relations, etc., can be found in any human language, though their superficial manifestation usually varies covering a broad spectrum; it is rather difficult on the other hand, to imagine a contextual situation of the external world that supports such phenomena. As a consequence, the explanations based on exaptation have to be ruled out.

2.3 The big mutation hypothesis

This is the reason why a much more radical explanation was tried: mutation. Pinker (1994) claims that humans share mentalese, a language acquisition device that allows them to learn languages and which they acquired starting from the communicative skills of their ancestors. Nonetheless, Pinker rejects the position of Deacon. The reason lies in the fact that Pinker holds Chomsky's innatist position, and therefore supposes that the linguistic features that humans share and that should have been acquired by means of Baldwinian evolution are those that characterize Universal Grammar, that is, a set of structural properties, like dependency relationships, or a set of grammatical categories, like tense and mood.

Deacon points out that it seems hard to imagine some natural context that preferred the inheritance of one version of a syntactic structure rather than another competing one. It stands to reason that the external world can only be responsible for some referential properties of language, such as noun and verb (the phenomenological counterparts of things and events respectively), or, as Deacon claims, the symbolic reference, whereas structural properties are absolutely excluded. According to Deacon's view, grammatical properties constantly vary from one language to another and they cannot become settled in the genome. However, as linguists know quite well, there are many invariable grammatical properties which can be found in any language.

These properties are very important in Chomsky's universal grammar hypothesis: the control theory, the trace and bounding theory, the binding theory, the alpha movement, the empty category principle, the pro-drop parameter, etc. (cfr. chapter V). No agreement between Biology and Linguistics or, more precisely, between Darwin and Chomsky, seems to be possible until a suitable explanation for the biological emergence of grammar is found. In fact, many scholars working inside the Chomskian paradigm maintain that such an explanation is very unlikely for the moment, a deep alteration of the state of art in Biology would be necessary to make this possible. It is no wonder that linguists have become reluctant to accept biological explanations of the emergence of language.

This is the position of Jenkins (2000), who allegedly expresses the Chomskian ortodox attitude, and not that of Pinker. Pinker (1994) argues that there was an important mutation that suddenly improved our language faculty. Humans had not acquired these grammatical properties in a context dependent way, simply took advantage when an evolutive leap directly encoded language in our genome as a neurologically hard-wired universal grammar. Perhaps, the so called 'hopeful monster story' finally came true!

Unfortunately, current advances in molecular Biology do not support his claim in any way. Mutations are caused either by normal cellular operations, or by random interactions with the environment. There are two classes of mutations: point mutation and insertion mutation. Point mutation changes only a single base pair as a result of an incorrect synthesis or of an incorrect replication or repair: it consists of a base mispairing, which is called transition when a pyrimidine or a purine is substituted by the other, and transversion when a pyrimidine is replaced by a purine or vice versa. Insertion mutation, on the contrary, is caused by transposable elements that are sequences of DNA capable of moving from one place to another.

Pinker does not explain what kind of mutation would have produced the emergence of language. As point mutation only causes slight deviations of regular behavior, it seems unlikely to have supported any linguistic skills because only humans are capable of speaking and speech cannot be related to the communicative abilities of any other animal species. Besides, there is no evidence for directed variation in mutation, it does not consistently produce changes in any particular direction. Humans are apes that have added a language to their incipient communicative skills, but theoretically there could also be apes that have partially lost this ability, or even humans that have lost some of the properties that characterize a language. Curiously enough, however, no intermediate steps are actually found in communication. The distance between the communicative patterns of apes and humans is a consistently insurmountable one.

On the contrary, if linguistic mutation has to be attributed to the insert mutation type, then the problem now is how do we explain the nature of these radical stretches of DNA that are responsible for our linguistic skills.

Were they already language specific and, if so, why didn't they initiate a communicative skill before moving elsewhere within the genome? And if not, that is, if language ability only arises when combining transfered material with the surrounding guest DNA, do the formal inner principles of grammar – the X-bar theory, the move α operation, and so on – result from a unique mutation, and was this responsible for the emergence of language as a whole?

This hardly seems conceivable, as any organic behaviour is due to many genes at the same time, and the more complex the behavior the more sophisticated the gene combination that supports it. In other words: the great mutation proposed by Pinker appears rather as a series of mutations, each guaranteeing a specific principle that follows a specific set of rules and categorizing strategies. However, such a progressive building up of linguistic ability contradicts once again the fact that no intermediate species between humans and the order of primates has been found. Theoretically, we could reasonably expect there to be a species with only one of the universal principles of language faculty (for example, X-bar theory), another species with two principles (say, X-bar theory and binding theory), and so on. When we look at the number of bones in a mammalian skeleton, from mice to humans, this is just the case. But language is human specific, and no gradual mutation seems to fill the gap between our species and our closer ancestors.

Further, current discoveries in Neurology (Rice and Mahwah, 1996) have made it clear that language disorders may be related to genetic impairments, but they always manifest through a wide range of disturbances that relate to several grammatical components at the same time. For example, some alterations of the q31 region of chromosome 7 are responsible for language disorders such as the Tourette syndrome or the autism. Disabilities related to dyslexia come from a disturbance in the region D6S464-D6S273 of chromosome 6p. The loss of speech in women, that characterizes the Rett syndrome, is related to a mutation of gene MECP2 in the region q28 of cromosome X (Stromswold, 2001). Nevertheless, this evidence does not support the existence of the 'genes of grammar' that were claimed for by Pinker. They only demonstrate that language is performed by very complex neural webs and that the failure of a protein, which is activated by a specific gene, may disturb the synapses within the web producing speech disorders. Whatever the internal plausibility of innatism, the idea of the gene(s) of language looks rather untenable.

2.4 In search of another explanation: emergence

As a matter of fact, Chomsky's argument, when he points out that children learn language very quickly from sparse and chaotic input, convincingly demonstrates that language ability is innate. No other animal species is capable of this. For the time being, no consistent theory has yet rejected the poverty of the stimulus hypothesis and its consequence, the genetic built-in faculty of language in humans. The difficulty, as Chomsky recognizes himself, lies in how we should shape such a claim in evolutive terms. We first adopted the starting point of the genetic foundations of grammar, and then we are upset when we find that no mechanism, neither Darwinian evolution, the Baldwin effect combined with exaptation, nor mutation, seems capable of providing an explanation.

But an inside out approach does not require language faculty to be a result of Darwinian evolution. Although many generativists support this view, Chomsky himself seems to prefer an explanation based on Physics (1986, 2002). The idea is that there is not anything like the genes of grammar, rather grammar emerged as a by product of the increased hominid brain. Language is a complex system and it could only come about on earth as a consequence of complexity. When several millions of neurons interact in a very narrow space, a new formal structure emerges: human language. Palaeontology registers show a growth of brain capacity from primates to humans. When a hominid's brain grew big enough, it suddenly became able to support not only the structural and functional necessities of communication by means of shouts and gestures, but also a new emergent system: the linguistic system. This system is a formal system whose properties were not shared with any previous communicative primate model.

It is interesting that Chomsky has strongly simplified the set of formal features of his universal grammar. In the P&P (principles and parameters) model (Chomsky, 1981) there were a set of properties which any grammarian, even a non generative one, would recognize as his/her object of study: hierarchical structure (X-bar syntax), order of morphemes and the changes that are allowed (move α), anaphora (empty categories and trace theory), etc. Scholars may disagree on the solution Chomsky proposes in every case, and they may think that other facts such as communicative functions should be considered also, but this is undoubtedly syntax.

In MP (minimalist program), that according to Chomsky (1996) should be considered as a program and not a model, the formal characteristics of language are drastically reduced. Actually, he only recognizes two: merge and move. It is clear that such a dramatic cut is related to the necessity of deriving formal syntax as a by product of complexity. Unfortunately, although Chomsky is probably right, these properties constitute a necessary condition of language, but they are not yet sufficient. As we will see, any perceptual grasping of the external world is also based on merge, and dynamic vision supposes move.

No doubt this proposal is a stimulating one. However, the theory of complexity, like any other mathematical model, is supposed to work in a similar way every time a similar set of initial conditions occurs. Unfortunately this is not the case when we compare language to other formal systems. As pointed out by Humboldt (1836), language makes infinite use of finite procedures. This property, the so called 'particulate principle of self diversifying systems' (Abler, 1989), should be biologically isolated (Chomsky, 1975, 2000) as soon as language is isolated too, but, as a matter of fact, it is a property of many other natural systems such as Chemistry, Physics, or Genetics. The question now is how Chomsky can explain the uniqueness of the formal properties of grammars which he described. If move-α, X-bar, empty categories and so on were formal consequences of the particulate principle, one wonders why they do not exist in Chemistry or in Mathematics at the same time.

3 Outside in: the evolution of language

We could, instead, consider the evolution of language as a central, non ephiphenomenic process, without postulating any necessary relationship with the brain. Brain and language would have evolved independently, until each was capable of fitting the structural complexity of the other. Of course, this implies considering language basically as a social phenomenon and, therefore, to abandon the Chomskian cognitive paradigm. This is the choice we are going to examine next. In supporting it, three difficulties arise: the way language is learned; the way a social calculus converted into language; the nature of language itself.

3.1 Learning a language

The idea that language is not biologically rooted and that the gap between humans and all other animals could simply be filled by considering our superiority in the learning process is, of course, a well established point of functional linguistics. A consequence of this would be to consider language among other cognitive skills. As stated by S. M. Lamb (1998, 375–376):

> The linguistic system is connected to other subsystems of the overall cognitive system [...] The cognitive system does not have nor does it need places to store symbols like those of taxonomic or generative linguistics, or of rule based artificial intelligence models. Since its information is in the connectivity of the network it requires no storage space other than the network itself [...] The linguistic subsystems have their basic structural properties in common with other cognitive systems, including hierarchical organization and similar nection structures and network operations.

Unfortunately functional considerations relative to this topic do not include the fact that language differs from any other cognitive skill since it is an unlearned faculty, it tends to appear around the age of two. The verification that humans manage to learn many things that animals cannot is an obvious one: we drive cars, we cook, we sew, we write, etc. These abilities are a consequence of brain development, particularly of the frontal lobe, and, hence, the result of the very complex network of neural connections at its disposal. Similarly, those scholars would argue that, language is learnt on the basis of an enlarged brain. The problem of such a hypothesis has always been the same: many humans learn to do anything they are taught, but only language is learned without apparent effort. In other words: language faculty seems to be innate in spite of everything. Children learn to speak in essentially the same way many animals – humans among them – learned to walk or to fly: with some help given by adults, but always with much less help than they would have needed if that ability had to be mastered from the beginning. No biologist has ever doubted that flying is an innate property of birds, although most birds usually receive some lessons from their parents. Similarly, language or, better said, the language acquisition device, *must* be an innate property of the human species, although children will only start to speak when the data of a specific language surround them.

3.2 Two outside in models: motor control and gesture

This does not imply that the outside in models of the origin of language are wrong. Until now we have examined some arguments which support the hypothesis that language appeared inside the hominid brain and later spread out. Now we will consider the opposite proposal which establishes that language is a matter of communication that first began in the social group of primates. The motor origin theory of language evolution is established this way. According to it, the evolutionary adaptation for language could be considered as the end-result of a series of evolutionary modifications to the serial motor control systems of the primate brain. There is some evidence to support such an idea: there is a link between handedness and language dominance, and there is a clinical correlation between aphasia and apraxia.

Lieberman (1975) and Kimura (1993) suggested that the serial-ordering capabilities of the primate motor system came under intentional control in hominids and evolved later to a sophisticated vocomotor system. Recent discoveries (Rizzolatti and others, 1995) of the so called mirror neurons in primate brain may support this view. Mirror neurons of apes fire not only when they are performing some activity, but also when they are seeing another ape which is performing it. And interestingly enough, mirror neurons are located in the F5 zone of the frontal lobe of the primate brain which corresponds to the Broca's area in humans!

Motor control represents a kind of formal protosyntax, but what about meanings? The semantic aspect of the outside in approach has been supported by some research into social cognition. Donald (1997) proposed a model where mimesis plays a central role. Mimesis is a non-verbal representational skill rooted in kinematic imagination. Humans have a set of mimetic abilities, such as iconic gesture, pantomime, imitation and the rehearsal of skills. But these abilities are also present in apes: Donald supposes that mimesis evolved in hominids and set the stage for the later evolution of language. Donald's idea seems conclusive not only because of the fact that humans use mime every time they are unable to be understood by other people, but because it establishes the fundamentals of intentional expression in hominids. Language could not have appeared suddenly, thus in order to arise by evolution a communicative environment was needed in which natural selection could act: the mimetic behavior of hominids could have represented just this cognitive layer.

Motor control theory on the origin of language and mimic theory are not contradictory. Neither do they contradict the pragmatic requirements that prove to be necessary for such a practice. For gestures to be adaptive, hominids had to live in society and possess hierarchically organized behaviour (Tomasello, 1999). It seems certain that they did. In order to understand others, hominids adapted an analogy with the self: knowing that someone else had psychological experiences like their own, they wanted to manipulate those states for various cooperative and competitive purposes. Notice that the theory of mind is a natural consequence of the evolution of mirror neurons.

The outside in hypothesis does not envisage language as an adaptation, but rather as a manifestation of more general abilities, such as general intelligence, which in its turn would be the adaptation. Grammatical structures had simply emerged from discourse patterns over historical time. The difficulty lies in the fact that we know quite well how grammatical structures arose in historical time, and the sequence is an unidirectional one, from lexical item to grammatical morpheme (Heine, 1991; Heine, Claudi and Hünnemeyer, 1991). It always happened that a full word was generalized and converted into a morpheme, never the other way round. This explains some grammatical regularities of world languages, but not the formal properties we are interested in, those that belong to universal grammar. We may expect that a lexical item that means space converts into a specific preposition, but no lexical origin can be imagined for hierarchy or for changes in order. It is a well known fact that cultures differ in relation to their degree of complexity, but that human languages do not: if these grammatical properties (hierarchical organization, move, empty categories, etc.) had been a consequence of the erosion of lexical items, one could expect that as more lexical items eroded so more grammatical structures were created. Nevertheless this did not happen.

3.3 The evolution of language culture

A turn round would be, as suggested above, to consider that language belongs entirely to culture, that it is neither innate nor a result of previous adaptations of the brain. This is the philological approach (as opposed to the linguistic one). It constitutes a radical point of view. Do not worry about the formal properties of language. Social groups interchange merchandises according to certain economic laws, and, although economy has a mathematical background, these laws do not fit the formal structure of the brain. There would be no sense in saying that economics is innate, although we know that children perform in economic terms as soon as they exchange some goods with their environment.

The philological – i.e. strictly cultural – approach has been predominant in the science of language for many centuries and it allows us to rule out the origin of the language problem as such. Notice that culture was born before language: the *Homo habilis* (2,000,000 years B.C.) had a manufacturing culture and even the *Australopitecus ghari* (2,500,000 years B.C.) seems to have had one too, but language, according to the less conservative estimations, is only 100,000 years old (*Homo sapiens*, and perhaps *Homo neanderthalensis*).

M. Christiansen and S. Kirby (2003) think that humans appear to be biologically adapted to language because language has culturally adapted to them. And they argue that humans can survive without language, whereas any language that was unable to adapt to the structure of our brain simply became extinct. Such a view is influenced by the difference established by R. Dawkins (1976) between *genes* and *memes*: genes are the units of natural evolution, the stretches of DNA that jump from one organism to the next in an evolutionary line; memes are the units of cultural evolution, information bits like political, aesthetic or religious ideas that also spread around in a social group by jumping from one mind to another.

The question now is that of the evolutionary nature of linguistic units: should they be like genes, or like memes (ideas), as is implied by the claim of Christiansen and Kirby? One could argue that linguistic units are like memes since both must be learned in a social group, they do not develop by themselves as organisms do. Nevertheless, at the same time it could be argued that they are not like memes, but rather the vehicle in which memes move from one person to another, and that in Dawkin's approach the organisms are also vehicles that allow the biological information to perpetuate by transporting a pack of genes during their life time. In other words, the form in which language unities are born demonstrates neither their biological or cultural nature.

Christiansen and Kirby also point out that languages evolve far faster than organisms. For example, Romance languages were derived from Latin in only five hundred years, whereas the descent of *Homo sapiens* from their ancestors took at least a million years. However, considerations based on the speed of the evolution look rather weak: similarly, there are organisms, like humans, that are replaced by their descendants in a period of half a century or more, whereas others, like bacteria, only take a few minutes to do so, and both are organisms. Furthermore, memes do not always change quickly. This is the case with things like clothes, books or even political ideas, all of which, it seems, are constantly in or out of fashion, but the taboo of incest or that of death, for example, have remained unchanged since humanity began.

Neither ontogeny nor speed throw any light on the subject. The question whether linguistic units are analogous to biological units or to cultural units can only be answered by considering the modalities of their evolution, that is, phylogeny. How did languages evolve?

3.4 An old fashioned comparison: language as an organism

There is no doubt that the proposal of reducing language to culture looks attractive. But we cannot support it before considering the nature of linguistic evolution compared with the evolution of the natural world. In the second half of the nineteenth century the ideas of Charles Darwin spread to many sciences, including Linguistics. Although his well known book of 1859, *On the origin of species*, only challenges the evolution of beings with the evolution of languages on four occasions (Bergounioux, 2002), several scholars took it seriously as a starting point for their own works: Withney in Great Britain, Schleicher in Germany, and Darmsteter in France may be considered 'biolinguists'. As pointed out by Schleicher (1873, 30):

> Languages are organisms of nature; they have never been directed by the will of man; they rose, and developed themselves according to definite laws; they grew old, and died out. They, too, are subject to that series of phenomena which we embrace under the name of "life". The science of language is consequently a natural science [...] The rules now, which Darwin lays down with regard to the species of animals and plants, are equally applicable to the organisms of languages, that is to say, as far as the main features are concerned.

Such ideas were induced by Darwin (1871, 465) himself when he judged "the formation of different languages and of distinct species [to be]" [...] curiously parallel". Unfortunately, this comparison was based on a rather loose analogy. It is possible to think up many artificial structures, besides language, that are born, that develop, and that come to an end with no human thought supporting their evolution. For example, axiomatic formal systems of Logic or Mathematics, which begin as a set of axioms, develop into a set of intermediate structures, and conclude as an inventory of laws. No wonder that the derivation of the syntactic structures of the language had been built up according to that picture many times.

It is true that natural languages, unlike those formal algorithms, are sensitive to human interactions, language is a social system. But, as F. de Saussure criticism on Schleicher's picture pointed out, any social system would remind us of an organism but this fact does not allow us to go any further. The expression "independent of human desires", as applied to language in Schleicher's sense, can only mean "not influenced by teleological goals". But in that sense, any social system has to be considered independent in as far as their actors interact mutually without having in mind the group goal. The scientific discredit of the biological metaphor in Linguistics at the end of the nineteenth century was closely related to the rejection of natural explanations in social sciences at the same time. It would be as unjustified and metaphorical to speak of the struggle of life by Spencer as to speak of the linguistic organisms that are alive or dead by Schleicher. No wonder that the Darwinian points of view never appeared in Linguistics again: current research that aims to reconcile Darwin with Chomsky (Pinker and Bloom, 1990:

Calvin and Bickerton, 2000, etc.) is intended to justify the innate character of language, but does not consider language as an organism in any way.

Nevertheless, comparisons drawn between human language and some biological structures are not always wrong. The problem is rather that of correctly extending analogy to homology. According to the Webster's dictionary, an *analogy* is a "correspondence in some respects between otherwise unlike things"; on the contrary, two or more beings are said to be *homologous* when they are "similar and related in structure, function, or evolutionary origin".

Analogical comparisons usually do not go further. For example, we can compare the classification of sciences with a tree, as Porphyry did in antiquity, and note that like the trunk of the tree which splits into two or more branches, a scientific subject can also divide into new lines or branches in the same manner. However, no extended similarities between trees and taxonomies are acceptable: the branches of a tree never meet again in nature, whereas two independent sciences that had a common source (say Chemistry and Biology, both coming from Aristotelian Natural Physics) can merge in some respect to produce a new science (Biochemistry).

On the contrary, an homology-like consideration is based on formal grounds: we can extend a set of properties from the formal component of the comparison until the substantial one. Recent developments in the morphodynamics of language have taken great advantage of the application of general topology to linguistic structures. For instance, I tried to predict the general linguistic levels and the formal categories of human language starting from general topological laws (López-García, 1990), while W. Wildgen (1999) has demonstrated that the morphodynamic model introduced by R. Thom (1972) may predict a varied set of linguistic processes such as the historical establishment of lexical units, the actantial patterns that are built up in the sentence, the modalities of word formation, or the constitution of discourses. However, no general dynamic model has yet been proposed in order to predict how the evolution of languages was able to take place.

By now, no matter how eccentric Schleicher's proposal may have seemed a century ago, neither the static comparison of topological structures in Biology and in Linguistics, nor their dynamic challenge are prohibited. In this vein, however, there is no reason to hypothesize about (as Schleicher did) a deep congruence between, on the one hand, biological evolution across simple-to-complex hierarchies of species and, on the other hand, linguistic evolution across simple-to-complex hierarchies of morphological types. And the question is: what was wrong with the Schleicherian approach?

3.5 A new comparative approach: language as a species

I think Schleicher failed to create a suitable basis for comparison. He supposed there are two formally similar sets, the organism, conceived of as a set of cells, and the language, that he considered to be a set of linguistic units (sentences, phrases, words, and so on). This is the reason he compared the radicals of the words to the nucleus of the cell and the affixes to the protoplasm, a descriptive analogy that lacks scientific power and that can only be employed for didactical purposes. But there is another possibility, supposing, as is the case, we consider that evolution primarily affects the *species* and not the organism. One single organism does not evolve, it matures: the concept of evolution implies that the descendants of that organism change slightly, and so on. Taking the species as a starting point in Biology sheds some light on the concept of language in Linguistics at the same time: from this point of view, a natural language – i.e. a linguistic species – can be considered as a set of speakers that mutually interact.

The basis of comparison of both topological spaces, the biological and the linguistic, would look as follows:

BIOLOGY	LINGUISTICS
A biologic species is a set of organisms that are able to maintain sexual interchanges. When they fail to do it, we say they belong to different species.	*A language is a set of speakers that are able to maintain verbal interchanges. When they fail to do it, we say they speak different languages.*

This starting point is by no means trivial. R. Dawkins (1976) has demonstrated that, in strict Darwinian orthodoxy, evolution should be attributed to genes rather than to organisms. Organisms die, genes survive. In order to do this, genes pass from an organism to its descendant and during this transition their genome varies slightly. Similarly, speakers die, but natural languages survive. As we have seen, Dawkins pointed out that the equivalent of genes in human sciences are memes, that is, cultural units that are transferred from one generation to the next via language. Nevertheless, there is a kind of psychological reality that is even closer to genes than memes are. Call *gene* a biologic propriety that may be inherited and call *lingueme* a linguistic feature that can be reproduced in the next speech act. This unit was coined by Croft (2000, 239) who defined it as follows: 'A unit of linguistic structure, as embodied in particular utterances, that can be inherited in replication; the replicator is the basic linguistic selection process; the linguistic equivalent of a gene'. We can conclude then that *the set of genes of a species is its genetic patrimony, and the set of linguemes of a language will be its linguistic inventory.* This is what Croft called the *lingueme pool*.

3.6 The formal properties of linguistic evolution

Linguistics and Biology are, then, two sciences interested in evolutionary patterns, and in both cases *evolution* is related to the concept of *variation*. Languages change because their units (be they phonological, morphological, syntactical or lexical) vary, and some allo-units (allophones, allomorphs, etc.) replace others in certain contexts over time. In Biology, the descendants of organisms also vary and natural selection imposes the survival of some alleles according to contextual (i.e., environmental) conditions. In fact, no B-evolution is possible without hereditary variation (that is without a gene manifesting as several alleles), neither is L-evolution possible without linguistic units varying in some allounits. The comparison of L-variation and B-variation may be described on several levels.

3.6.1 The structure of evolution

From a dynamic point of view, every linguistic interchange would be equivalent to a biologic sexual interchange. Organisms vary because of recombination, that is, because one parent's DNA string does not exactly correspond to the other parent's in the genome of the descendant. Languages vary for the same reason, because the listener's understanding of the speaker's utterance does not equate with what was intended. As Bloomfield (1933, §20.10) pointed out: 'Historically, we picture phonetic change as a gradual favoring of some non-distinctive variants and disfavoring others'.

Variation is not always a gradual process. A second source of B-variation is mutation, an occasional change in the sequence of genomic DNA, and its L-variation equivalent would be sporadic linguistic change, that is, a spontaneous, slight change in sound (or meaning) on the part of the speaker may produce mutation by deletion (apocope, syncope), by mutual interchange (metathesis) or by insertion (epenthesis). B-mutation may be due not only to gene mutation but also to chromosome mutation. This kind of change (where part of an entire chromosome is deleted, inserted, duplicated, moved, or has collapsed with another part of a chromosome) bears a strong resemblance to the borrowing of lexical items from other languages. We can conclude that B-evolution and L-evolution are structurally homologous.

3.6.2 Methodology: samples for the study of evolution

On the other hand, B-evolution and L-evolution also overlap when we consider the nature of the samples usually collected by researchers of the two processes. Those biological evolutionary patterns that are congruent with corresponding linguistic patterns – on the basis of the formal properties they share in common – should be taken into account. The empirical evidences of B-evolution and the empirical evidence of L-evolution are similar: a) the fossils of a species are like the forms of the old texts of a language; b) the anatomic similarities between species play the same role as the typological similarities do in comparative linguistics; c) embryonic similarities of species remind us of the structural features that Creole languages share in common on the basis of their pidgin genetic origin (Taylor, 1971). We could add to the previously established structural homology between Biology and Linguistics in relation to evolution, a methodological homology at the same time.

3.6.3 The causes of evolution

The question now is whether L-evolution may be said to be due to a selective process like natural selection, that is, whether they are even causally homologous. Current research in linguistics does not accept the relativist hypothesis of Sapir-Whorf, that is, the adaptation of language to the conditions of the external world and the corresponding belief that some languages fit better than others to environmentally imposed conditions. Neither is there a direct relationship between biologic efficacy and adaptation in nature. Natural selection in B-evolution is a mathematical stochastic concept. There is a correspondence between natural selection and density which results in the less frequent alleles imposing themselves over the more frequent ones. Similarly, in L-evolution, educated variants, although less frequent than ordinary ones, usually survive.

Anyway, the typology of natural selection and the typology of linguistic selection are parallel. Natural selection may be: 1) Steady (varieties concentrate on medium scores); 2) Directional (medium scores move in a specific direction); 3) Diversifying (medium scores move in two directions). It is easy to see that L-evolution also behaves in this manner: 1) ritualistic languages, like classic Arabic, the language of the Koran, or Sanskrit, only preserve regular features and become fossilised languages; 2) linguistic change usually follows some direction, like the series of related changes that characterize the evolution of romance consonants; 3) sometimes a language is broken down in two different modalities, on the basis of geographical differences (British English vs. American English), or on the basis of social differences.

The most important functional consequence of evolution, either L- or B-, is species building, the birth of a new species. There is no agreement among scientists on the causes of biological speciation, nor do linguists agree on the processes that cause some languages to develop from other languages. The Darwinian approach considers speciation as a consequence of very gradual changes due to the action of natural selection on variation. There are other biologists (Gould, 1977), however, who have proposed a theory of punctuated equilibrium consisting of fast changes that create every species and that are followed by long (basically) static periods of time in which morphological features hardly ever evolve. On the linguistic side, we find this picture repeated. Although linguistic change is supposed to be gradual (Chambers, 2002, 364):

> When sociolinguists began viewing language changes in their social contexts, many of the old mysteries of historical linguistics [...] simply disappeared. For instance, linguists long recognized that rates of change fluctuate, and that periods of relative stability can be followed by periods of considerable flux. In times of flux, if change is viewed from discrete points in time rather than along the whole range, it can take on the appearance of a generation gap or even a communication breakdown.

There are some processes, like the sudden emergence of romance syntax in the Latin of Vulgata (López-García, 2000), or that of a new orthography and pronunciation in Carolingian texts (Wright, 1982), that have a catastrophic character. The opposition "gradual vs. punctuated" in the study of evolution is not only a result of scientific perspective. In fact, punctuated evolution does exist, and the sudden rise of entirely new ortographic or discursive norms has the same empirical value in linguistic evolution as discontinuous fossil evidence in Biology.

3.7 Linguistic speciation

3.7.1 Mechanisms of speciation

Biological speciation always implies the existence of some reproductive isolation, which can be catalysed by two kinds of mechanisms: pre-zygotic and post-zygotic mechanisms. Pre-zygotic mechanisms are due to: ecological isolation (each species fills a different ecological gap); seasonal isolation (each species blooms in a different season); behavioral isolation (no sexual interchanges happen between either species); physical isolation (sexual interchange is prevented because of the incompatibility of the sexual organs of both species). Post-zygotic mechanisms characterize the initial stages of a process of speciation, and consist of the inviability of the hybrids resulting from sexual interchange between both species, be it due to their inviability or to their sterility.

The speech situations that give rise to a linguistic speciation are not far from these shown above. The origin of such a process can be the *ecological isolation*, as when ecclesiastical Latin was only spoken in churches or monasteries, and it progressively diverged from the Vulgar Latin people spoke in the streets. On the other hand, sometimes we find a situation very close to *seasonal isolation*, as when a new species arises because the younger generation employs a vocabulary that the elder one does not understand: this is the reason why English was split into two different languages, old English and middle English, as a consequence of the Norman invasion. *Behavioral isolation* in speech communities characterizes those linguistic changes that are gender framed: the linguistic features that characterize the speech of each sex usually do not impede their mutual interchange, they only complicate it; but in extreme situations, as on the famous island where women speak Arawakan and men speak Caribbean, it does. *Physical isolation* can also cause linguistic speciation: this is the case with deaf people that are unable to speak oral languages and that, as recently shown by Kegl, Senghas and Coppola (2001), gave rise to a new sign language in Nicaragua. The most common situation in the linguistic variation processes that give rise to a new species is *geographical isolation* and, similarly, Biogeography demonstrates that any isolated microcosmos creates a great diversity of species: in this sense, the spread of finches in the Galapagos islands, as described by Darwin, looks like the incredible variety of languages in Australia.

3.7.2 Patterns and rythms of speciation

There are also striking similarities between the patterns and the rhythms of L- and B-evolution. A B-evolutive process can take place either inside a lineage (anagenesis) or inside-out it producing a lineage-fork (cladogenesis). Linguistic changes resemble this picture as they substitute an old system with a newer one (say XII century Spanish by XIII century Spanish), or one language by other(s) (Latin by Romance languages).

Moreover, several B-evolutions can develop in parallel because they share a common ancestor or a similar function: the similarities we discover among the extremities of mammals like dogs, humans and whales are due to their predecessors; on the contrary, the fact that bees and birds both have very similar wings should be explained because they are flying animals. It is easy to see that L-evolution knows both evolutive patterns. All Semitic languages – Arabic, Hebrew, Amharic, etc. – are homologous and exhibit irregular plurals or radicals made out of three consonants because these features were already present in the primitive Semitic. On the other hand, most modern languages, no matter where they come from, develop in a similar way in order to incorporate the new lexical roots that scientific discoveries were providing. Sometimes, B-similarities are called serial ones, as when there are several related organs, like the arms, the legs and the spine of vertebrates, which evolve in a parallel manner. But such a correlation is quite usual in L-evolution too: Grimm's law stated that primitive Indo-European unvoiced stops [p, t, k] changed in the Germanic languages to unvoiced spirants [f, th, h] whereas primitive Indo-European voiced stops [b, d, g] changed in the Germanic languages to unvoiced stops [p, t, k] and the voiced aspirate stops [bh, dh, gh] of the former change into voiced stops or spirants [b, d, g] of the latter, in which a serial homological process may be considered.

3.8 Language is culture but grammatical units are not cultural ones

We may conclude that B-evolution and L-evolution are homologous on a structural, methodological, causal, and functional basis at the same time. They share many features and patterns, and an extended comparison of both evolutive processes has proved worthwhile. The question now remains whether the similarities that result from such an interdisciplinary approach are merely due to the formal nature of both processes or a consequence of some deeper affinities between them. I'd like to pay close attention to the fact that, when applying the comparative frame to Dawkins' memes, many of the correlations stated above do fail in this case. For instance, human groups that originated in a common forebear group never evolve along similar structural patterns, but according to their individual environmental conditions. This is the reason why although Indo-European languages still look rather alike, Indo-European nations are very different from each other, as the British are compared to Iranians, or as the Spanish clearly demonstrate when compared to Hindus. Linguistics is not a natural science, as Schleicher alleged, but it behaves more similarly in some respects to Biology than any other cultural sciences. And the study of linguistic evolution will surely benefit from this fact.

When claims are made about the cultural nature of language, they usually suggest much more than they set out to do at first. Language is a product of culture, of course. Tool-making is also a product of culture. Hence, animals, which are only capable of making very poor tools, dispose of very rough communicative skills at the same time. The conclusion scholars commonly draw from such a parallelism reads as follows (Dobzhansky, 1976, 452):

> There is no doubt that communication by language, and the genes that made it possible, were promoted by natural selection. The cognitive structures on which both language and tool-making depend are the foundations of the control of the environment by culture. While all organisms adapt to their environments by changing their genes, man alone adapts mainly, though not exclusively, by creating the environments that suit his genes.

This is the argument that Christiansen and Kirby put in the same way twenty five years later as seen above. Unfortunately, Dobzhansky's tenet, one of the founding fathers of the synthesis paradigm in Biology, loses its impressive force when their successors extend it even to further developments of language. Language, like tools, is a creation of culture, Christiansen and Kirby say, the laws of natural selection cannot influence it. But language does not change as culture does, language speciation follows essentially the same pattern as nature speciation. This was the case with Cavalli-Sforza's proposal on the classification of the languages of the world which overlaps the classification of human groups: Cavalli-Sforza (1996) has demonstrated that the degree of genetic specificity of a human group correlates with its degree of linguistic specificity because both characteristics depend on environmental isolation.

Why do living things apparently become recognizable units called species, and how can one species split in two? This is a question that, as we have seen, will be answered in a very similar way by Biology and by Linguistics, but not by Engineering. Compared to language, the modalities of tool pattern splitting in two species are explicable only according to the necessities of a specific culture, and never follow predictable formal laws. This is true for any kind of tools, be they material tools or ideological ones.

Would such a statement mean that the evolution of language remains unconcerned about culture? It depends upon what we call a language. Cultural events only affect language at the lexical and phonological level, at the interface where it joins the world. A change in the social conditions of a human group – say, a revolution – or a change in life style – the conversion of a peasant society into an industrial one – has a great influence on the vocabulary. A change in the normal pronunciation group members listen to every day – perhaps as a result of the code switching that characterizes language contact – also changes their own phonetic habits. By contrast, grammatical changes are independent of cultural changes: modern technological culture is essentially the same in every country of the first world, but people still speak languages whose grammars differ as much as English, Japanese, or Chinese do.

As a consequence of this, grammar remains as unexplored from the point of view of evolution as it ever was. Grammar is related to information. An increase in the ability to gather and process information about the environment puts humans above animals, animals above plants, and plants above bacteria: in this sense, mankind has arrived at the top of the evolution tree. Nevertheless, all languages share the same set of deep grammatical features and the reason seems to be that they are unaffected not only by cultural changes, but also by biological changes. In other words, we have to explain the surprising fact that a property of human mind that genes made possible by natural selection – either inside out, or outside in – no longer evolved after humans appeared on earth. This is the challenge we will face up to in the following chapters. The answer probably lies in a surprising property, that memes and genes share in common, but which to my knowledge has never been pointed out. It is that: *although genes certainly change every so often, the genetic code never does; although linguistic texts, that support most memes, also change, the linguistic code does not evolve either.*

4 Formal inheritance

4.1 Do formal patterns precede their manifestation?

The hypothesis I will consider is that language has been formally built in accordance with the genetic code itself, no matter whether it evolved outside-in or inside-out. But this, unfortunately, is a conclusion we ought not to derive unless, of course, we wish to be guilty of methodological heresy.

This kind of heresy is known as 'transcendental morphology', a constraint on natural selection that the model of the synthesis (the model that reconciles Darwin with Mendel) has always vigorously rejected. In Darwinian terms, organisms evolve randomly without any formal pattern directing the pathways they adopt during evolution. Natural selection would allow only those variations that adapt better to the environment, and the only constraints that such a process will be exposed to are precisely the environmental changes. However, since the publication of Darwin's main book on evolution, and even long before, several scholars made substantial claims against this conception. I will rapidly examine three of them: the leaf archetype of W. Goethe, Geoffroy's idea of the vertebral ground-plan of antero-posterior differentiation, and the theory of form by D'Arcy Thompson. The reason for selecting them is that they illustrate three different approaches to the heresy: 1) The claim that biological form is not the result of some unpredictable interactions within the laws of nature, but that there is a previous abstract pattern, the archetype, to which those forms have to adapt (Goethe); 2) the assumption that there is no direct relationship between body organs and their functions, and that they behave relatively independently of the environment (Geoffroy); 3) the idea that nature is the domain of necessity, and that animate (biological) structures, like inanimate ones, are committed to physical forces that follow mathematical laws (D'Arcy Thompson). Although each position emphasizes a different aspect of the method, the point that all 3 share in common is the thought that evolution does not proceed at random, but according to a more general pattern.

Goethe developed the theory of archetypal morphology in his 1790 paper, *Versuch die Metamorphose der Pflanzen zu erklären*. There he proposed a single archetypal form for all plant appendixes, which predicts both the boundaries and the possibilities of the realized leaf forms. The central idea is that the formal leaf represents a pattern for all plants growing from the central stem, be they cotyledons, petals, sepals, leaves proper, or fruit. The procedure Goethe followed to reach this conclusion was to compare the apparently diversified structures of the vegetal world with each other. Faced with the clear diversity of the empirical data, and the overwhelming variety of vegetal organs, he postulated deflecting forces to justify discontinuities, such as the cycles of expansion and contraction. His formalistic commitment opposed Goethe's system to the primacy of adaptation in Darwinian terms, although he admitted it as a complementary, although always secondary, force. In Goethe's view it is because a plant has been shaped from within that it succeeds in adaptation, and never the contrary. This is a typical Platonic view: pure ideas precede their manifestations in the world; as a consequence of this, ideas exist in their own right. Little wonder that Darwin and his followers rejected Goethe's position: since the scientific method was postulated in the XVI century, science has relied on verifiable facts, not on ideas.

Geoffroy de Saint Hilaire developed a similar view on animals in several papers he wrote during the first half of the nineteenth century. But, compared to Goethe, he was rather Aristotelian. He was opposed to the functionalist position of Cuvier which stated that nature works constantly with the same materials, and causes the same elements to reappear, in the same number, in the same circumstances, and with identical connections. Working on bones Geoffroy discovered a formal homology of the bones of the shoulder griddle in birds and, surprisingly, in fishes too. In a further generalization of the theory, he established the common structure of vertebrate and arthropod segmentation. Geoffroy was aware of the fact that both phyla manifest inverted orientation: the single nerve tube of vertebrates runs along the dorsal surface, whereas the two main nerve cords of arthropods run along the ventral surface. The solution he arrived at was to postulate the same ground-plan with reversed orientation, viewing arthropods as vertebrates turned on their back. Drawing on this conclusion, Geoffroy argued that the functionalist credo that organs must exist in animals for a specific function (e.g., the wishbone in birds for flight) is meaningless. Notice that this theory comes closer to Darwinian claims than Goethe's when it points to the evolution and the adaptation of body organs to the environment. Both also share the rejection of the teleological argument that evolution has a particular goal. However, the distinction made by Aristotle between form and matter (the so called hylomorphic theory), is supported by Geoffroy, not by Darwin: the programme of the synthesis absolutely excludes the idea that an abstract implicit form can direct evolution.

D'Arcy W. Thompsons' theory of form was written back in 1917, *On Growth and Form*. The basic idea is that there are physical principles that build good forms through the direct action of physical laws on plastic material. Notice that such a theory is very far from the *Naturphilosophie* and its idea of a *Bauplan* in Goethe's, and even in Geoffroy's conception. According to D'Arcy Thompson, the forms adopted by living beings are neither mysterious nor ideal. Just as when a specific force that is applied to a stone moves it a specific length at a specific speed, some specific forces acting on the ontogenesis of a living being determine its particular form. It is all manoeuvred by Physics and nothing is owed to Philosophy: in fact, mechanical optimality is the unique criterion for his claim. But, then again, neither does Thomson's paradigm satisfy the requirements of Darwin's theory. The problem is that the geometrical transformations postulated by the former lack continuity while saltation constitutes an explicit argument against the latter. In other words: the forms of the theory are discrete, passing from one form to another without intermediate steps. Bateson (1913, 36–37), a geneticist directly influenced by D'Arcy Thompson, has applied these ideas to Biology and has also demonstrated the challenge they represent for the current Darwinian paradigm:

> It is in the geometrical phenomena of life that the most hopeful field for the introduction of mathematics will be found. If anyone should compare one of our animal patterns, say to that of a zebra's hide, with patterns known to be of a purely mechanical production, he will need no convincing that there must be an essential similarity between the processes by which the two kinds of patterns were made [...] When the essential analogy between these various classes of phenomena is perceived, no one can be astonished at, or reluctant to admit, the reality of discontinuity in variation [...] Biologists have felt it easier to conceive the evolution of a striped animal like a zebra from a self-coloured type like a horse [...] as a process involving many intergradational steps; but as far as the pattern is concerned, the change may have been decided by a single event, just as the multitudinous and ordered rippling of a beach may be created or obliterated by one single tide.

4.2 The inheritance of form in the natural world

The three heretical claims we have just examined above do not intend to justify the inheritance of form in the sense that forms could be transmitted from one species to another, bearing in mind that within a given species the infants always inherit their body forms from their parents. According to Goethe, form is simply there, as an abstract constraint, and beings are forced to adapt to it. In Geoffroy's view, formal patterns result in comparing several beings with each other, but he never argued that form may be passed on from ancestors to descendants because this would require the maintenance of form, rejecting thus Cuvier's argument. The third position we have examined is that of D'Arcy Thompson, he postulates the evolution of form, but not necessarily its inheritance, he points out that there are physical powers that conform the organs in every case. Surprisingly enough, however, recently geneticists have described a phenomenon that fits the epistemic gap we are now looking for: a formal pattern that exists in nature and that some organisms have inherited; the homeobox (usually abbreviated as hox).

The history of the homeobox hypothesis starts with the geneticist we referred to in the last paragraph, William Bateson. Bateson observed that there are some mutations with the peculiar effect of causing the typical form of one member in a serial array to develop in a different location usually occupied by another member of the series: for example, flies antennapedia, which have legs where antennae ought to be, etc. The next step is represented by Edward Lewis who explained the *Drosophila bithoraxoid*, an example of a fly that has an extra pair of legs: the bithoraxoid complex had evolved by gene duplication, with all members remaining aligned in a tandem array of eight genes on the third chromosome; each gene turns on in sequence with expression in successively more posterior parts of the developing larva; those mutations that induce a loss of function should weaken the gradient and cause anterior structures to develop in a more posterior position, for example, legs in abdominal segments like *Drosophila bithoraxoid*. Most of Lewis's original points have recently been refined, but the fundamental fact remains: there are some topological patterns of the chromosome that are responsible for some parallel structures of the organism with a correspondence relating to both tandem arrays. For instance, Lewis proposed the following scheme:

H T1 T2 T3 A1 A2 A3 A4 A5 A6 A7 A8   Larval segments of Drosophila
g1 g2 g3 g4 g5 g6 g7 g8 g9 g10 g11 g12 BX-C complex of genes

where H, T1,2,3, and A1,2,3,4,5,6,7,8 are successive segments of the body of Drosophila, from the Head through to the Thorax to the Abdomen, and g1...g12 are a dozen of genes that correspond biunivocally to them and that are arranged exactly in the same order.

Now we know that close to bithorax there are nearly a dozen genes responsible for the segmentation of Drosophila, as claimed by Lewis (Ubx, abx, bx, bdx, pbx, iab2, iab3, iab4, iab5, iab6, iab7, iab8), but only three hox genes (Ubx, abdA, AbdB), since several of Lewis' genes belong, as regulating segments, to a single gen. Besides, there are another six homeobox genes located in a different block. This eliminates the one-to-one (and somewhat mechanical) correspondence between body segments and genes, although it leaves unaltered the fact of linearity, the first hox gene Ubx controlling the front segments of the body, the second gene abdA controlling the intermediate segments, and the third gene AbdB controlling the back segments.

But the outstanding discoveries made by Molecular Genetics in the last twenty years permits us to go a step further. When geneticists read the DNA composition of the homeobox genes of many animals, they discovered that:

i) Homeobox appears in every animal species;
ii) Animal species possessing bilateral symmetry (bilaterians) have about ten hox genes;
iii) Hox genes are lined up in the chromosome;
iv) Mutations that alter the expression of a hox gene of a bilaterian animal may be restored if this gene is replaced by the corresponding homeobox of another animal. For example, the third homeobox gene of a mouse may restore the damaged third homeobox gene of a bee. This means that a specific homeobox gene is not directly responsible for the segment it is associated with: the gene does not make it, the gene simply regulates the expression of many other genes that participate in that task.

What kind of formal structure are we faced with when considering the hox genes? The answer to this question partially legitimizes the morphological heresy we considered above: the hox series is obviously a formal topological pattern that any species inherits from its ancestor for homeobox to be present in every animal. But homeobox is neither an abstract idea, nor the result of the combined effect of some blind laws of physical nature: hox genes are simply a stretch of 180 base pairs of DNA, – in other words, they are biochemical compounds.

These considerations do not imply that homeobox is immune to evolution. Although bilaterians have about ten box genes, as we said, their precursors, that already lived before the so called Cambrian explosion took place, have only three or four, and, more important, do not show the typical organization of co-linearity that defines the formal pattern of bilaterians. For example, Martínez et al. (1998) found that cnidaria, which possess radial symmetry, had some hox genes which, through duplication, gave rise to the bilaterian corresponding series. After taking such facts in consideration we are allowed to draw from them three fundamental conclusions:

First, form is inheritable in almost exactly the same way as when we say that matter – i.e. specific DNA sequences – is inheritable;
Second, the formal patterns that jump from every generation to the next one remain essentially unchanged through time, but, they are also subject to evolution in some cases;
Third, mature formal patterns constitute a colinear grouping of DNA sequences, that is, they have a topological identity; however, at the beginning, they were scattered sets of regulator genes.

4.3 The genes of language?

The difficulties that have arisen when considering the formal inheritance problem demonstrate that the genetic basis of the emergence of language remains largely unknown. Biological processes are not induced by genes in relation to the function they are supposed to perform. Genes may only regulate the expression of proteins, and every functional process comes from the combined simultaneous action of many proteins. This is the reason why we cannot speak of the gene for growth, of the gene for blood pressure, or the gene for anxiety: to a lesser extent we could speak of the genes for language. Some scholars have tried to evade this blind alley by claiming that linguistic genes are genes that determine a sample of proteins responsible for the constitution of those brain structures that intervene in language processes. The innate character of some formal structures relative to language knowledge would rely then on these genes.

The conclusion is probably true, but it looks rather tautological: any mental process is performed by the brain; language is a mental process; then, the genes of language are the genes that are involved in that mental process. In a similar way, religious beliefs are mental processes, and also mathematical thinking is a mental process, so both directly relate to their respective genes. As a matter of fact, there is very poor evidence to support the claim that specific gene impairments influence verbal behaviour. It is quite true that some recent discoveries carefully describe the case of families where specific verbal pathologies are produced by a local genetic default (Gopnik, 1990; Stromswold, 2001, for an updated review). But this only demonstrates that the neural structures involved in the expression of those linguistic structures are damaged. The tautology continues. As pointed out by Lorenzo & Longa (2003, 645–646) who comment on the discovery of FOXP2, a gene whose mutation provokes SLI (Specific Language Impairment):

> This finding can be considered as good news for the defenders of the biolinguistic approach to language, using Jenkins' (2000) label. It seems to strengthen the belief that language is biologically specified by one or more genetic factors, which in turn seems to fulfil Jerne's conditional respecting the integration of linguistics within biology. However, it is also a finding somewhat upsetting for linguists, because the aspects of language whose development seem to be under control of FOXP2 constitute, as we have noted, a heterogeneous class with no clear correspondence with any single level of analysis or domain of rule application ever proposed by theoretical linguistics. We thus face a situation in which, on the one hand, recent discoveries in the field of genetics seem to support the biological approach to human language, but, on the other hand, the results independently reached by linguistics and biology do not seem to fit each other.

4.4 On pregnance and salience: the concept of 'figurative effect'

The time has come to search for a different approach. Language is a form and humans do not acquire this form either as a consequence of learning only, or as a natural development of their mental faculties without any contextual input. The emergence of language is the result of combining inner and outer forces, of innate predispositions and of contextual stimuli at the same time. We are faced not only with the inheritance of form, but also with the way in which it was born. What we are looking for then, is, a dynamic theory of the birth and death of forms.

This theory was developed twenty five years ago by the French mathematician René Thom (1988). René Thom distinguished two fundamental concepts that are involved in the perception of forms: the salience and the pregnance. The salience is related to discontinuity: a form is said to be a salient form when it stands out against the ground. A noise that breaks the experience of the continuity of time is a salient form. An object that we perceive clearly in contrast with the ground space it comes from is a salient form.

Salient forms may affect our perceptual organs, although generally this effect disappears within a short time. They leave a trace in our memory, which also vanishes rapidly. But this is not the case with some forms that impress a biological meaning upon the perceptual organs of superior animals: for example, a predator's sight or smell of its prey. The recognition of these forms determines a strong response in the subject: an intense emotion, physiological reactions, and so on. These forms, that are also salient forms, will be called pregnant forms and pregnance the phenomenon they are responsible for as such. Pregnances act upon salient forms and modify them: the resulting state is a figurative effect.

Despite pregnances determining the entire behaviour of superior animals, there are very few that may affect them: hunger, sexual desire, fear. On the contrary, human life is entirely determined by pregnances. Thom thinks that every concept, which is manifested as a lexical item, is a salient form endowed with a pregnant potential. The boundary an infant surpasses when they leave their baby state to enter the child state is related to pregnance. As a baby, there were only pregnances like the hunger induced by the sight of food, a salient form, or the fear caused by hearing a noise: no fundamental differences exist between humans and animals at this level. But after the age of two, the words that surround children suddenly acquire a meaning for them, that is, a pregnant value: *mom, dad, water, car*, etc., at the beginning of their social life, *weekend, and much later freedom, unemployment*, etc., are like whirlpools that attract them and that modify their equilibrium very deeply. The gap that separates animals from humans is biologically insignificant, but from the point of view of pregnances it seems insurmountable: the former have no more than a dozen pregnances; the later live in a world made out of words, thousands of words, that is, thousands of pregnant forms, which are responsible for thousands of pregnances.

To understand how mankind crossed that line it is necessary to introduce the concept of 'cathexis'. If you remember the experiment by Pavlov with the famous dog, the sound of the bell that was associated with the piece of meat is said to be a salient form that was cathecticized by the food-like pregnance of the meat. Let B be the sound of bell, and A the meat; then, B > A, which means that B is the sign of A. Similarly, in the linguistic sign, a *word* is catechized by the REFERENT, that is, *house* > HOUSE. Words are, then, figurative effects which their respective meanings have induced, as pregnances, in the salient brains of the listeners that receive them. Once the figurative effect is built up, its form is converted into a source of spreading pregnance.

Pregnant categories may be excited by many stimuli, but some of them are primary ones. The source form of a pregnant category Cp is a form that is directly cathecticized by the pregnance, that is, a S such that there is no intermediate T where S > T > P. For example, a noise would be an intermediate form that produces the pregnance of fear in preys, but the visual form of a predator is a source form. Now the question arises whether the formal structure of a pregnant category may be genetically printed. In the case of superior animals, Thom says, the answer is negative: although it is easy to imagine the existence of innate smell signals, a visual form has so many contours in the three-dimensional space that no DNA sequence seems capable of codifying all the information it bears. Visual forms, like words, are source forms that may be transmitted only by culture, they are memes.

4.5 Linguistic pregnances: the emergence of syntactic laws

There is, however, the surprising phenomenon that biologists call imprinting: geese would rather hatch a rugby ball than their own eggs because they have a general oval form genetically imprinted, not a specific egg, and the bigger the 'egg' they find – the more likely they are to choose to incubate it. Thom concludes from this that in a pregnant category genetics may not program their salient forms but what he calls the topological structure of its potential pool. The emergence of human language is explained in a similar way: at the beginning the infant is entirely catechized by the pregnance of the mother's body (the baby is breast-fed, it looks at her, etc.), but later they pay close attention to the external objects, and the potential pool bound to the maternal pregnance gets subdivided into a lot of meanings. The next step is the emergence of syntactic structures: for example, in the transitive clause the pregnance emitted by the subject towards the object crashes into the individual pregnance of the object, and the result is the meaning of a verb.

These ideas belong to a scientific tradition quite different from the theory of generative grammar: in fact, Thom was the founder of a branch of cognitivism that never shared any assumption with the generativism, and that has remained hostile to it. However, it is interesting to notice how close their respective approaches are in relation to the subject we are interested in. Both Chomsky and Thom support the idea that humans develop some formal patterns that guarantee lexical meanings to be joined together in broader structures. According to Thom, the four fundamental patterns that organize the morphogenesis of biological forms look like this:

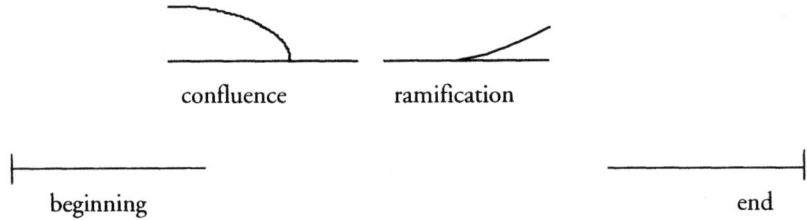

they are probably also the basic patterns that organize the minimal two-word structures in infant's speech. According to Chomsky, the properties of the computational system are merge, agree, and pied-piping. At first sight both formal patterning schemes have nothing in common, of course. But in a more accurate approach we find some deep resemblances between the topological morphogenetic structures, and the linguistic principles:

Merge is conceived of as a principle that determines how two lexical categories are joined together in order to form a new one that behaves formally as one of them. *Confluence* is a topological singularity consisting of two trajectories that meet, and later continue to move in the direction of one of them.

Agree is conceived of as a principle that guarantees the diffusion of a formal property starting from a category without leaving it until it reaches other categories into a specific domain. *Ramification* is a topological singularity that consists of a trajectory that bifurcates in two branches, one of them following the direction of the trajectory.

Pied-piping is the principle that marks a specific category as a chain that will be moved backwards leaving a trace after it. Its topological counterparts are *beginning* and *end*, that is, the birth of the category in a new place of the sequence and its death in another (|___ <.........< ___|).

The question now is whether these three patterns are genetically built in or if they emerge as a consequence of the interactions among neural cells. Notice that all of them are symmetry breaking properties: the two syntactical objects a and b that merge in a broader unit K(a, b) are differentiated from each other for only one of them projecting its syntactic category into the whole; similarly, agree relates a feature R, that is manifested in a to a lexical item b which is present in its domain; moreover, in the movement implied by pied-piping, there is a position where the lexical entity disappears (goal) and a position where it finally lands (probe). It seems as if the simple summing up of two symmetrical units in a compounding whole has given rise to a new property that was absent from their individual parts: the asymmetry that leaves operative the categorical behaviour of one of the units only and eliminates the other. The above properties could be considered, thus, emergent properties at first glance. Emergence manifests in any domain of science, from quantum physics to society.

However, every time a particular phenomenon has been reduced to a lower level of description, emergency as an explanation has to be ruled out. This could be the case now. Gestalt psychologists observed that perceptual patterns result from joining several stimuli (be they visual or acoustic) together. They established three laws that collaborate in setting the stimuli out in front of other stimuli that remain in the background:

The *closure* law states that stimuli tend to organize themselves in closed sets: it defines, then, a phenomenon that reminds us of linguistic government.

The *equality* law states that equal stimuli tend to group together: it defines, then, a phenomenon that reminds us of linguistic agreement.

The *proximity* law states that near stimuli tend to be grouped together: it defines, then, a phenomenon that reminds us of linguistic movement (for example when a relative pronoun approaches its antecedent: *the place [I lived in the place]* > *the place where I lived in*).

It is difficult to claim that perceptual laws are not inherited, at least partially, because they manifest as soon as an infant perceives a salient object within the surrounding ground: this happens during the very first days of life in the case of visual stimuli, and even before birth in the case of acoustic signals. Moreover, perceptual laws of mammals are simply the last stage or the end of an evolutionary line, related to the transmission of information packs, which began with the signals micro-organisms receive from the environment. The way genes work in the case of these perceptual patterns may be thought of as the cooperative interaction of gene activities and the global spatio-temporal dynamic of a hierarchically organized system.

However one shouldn't conclude that genes group the neural cells of the retina together as if they were a section mechanism. Gestalt laws are abstract formulations that summarize the resulting effect of genetic instructions, and not the way they actually operate. Genes simply contribute to the stabilization of cell groups into coherent clusters, they do not have to generate differences between groups of cells. We do not know exactly how our visual neurons work, but we can get an idea of perception from the findings about the olfactory bulb of rabbits by Freeman (1991). He supports the view that perception cannot only be understood by examining properties of individual neurons, which are guided by genes, but it depends on the cooperative activity of millions of neurons. Even so, the ability to detect up to 10,000 different smells is rather directly encoded in the human genome by the activity of up to 1,000 genes (Buck and Axel, 1991).

The human sense of vision differs in character from smell. Smell depends on an exhaustive alphabet of 10, 000 units. Vision, even if we adopt 10,000 distinct colours, is still faced with many unresolved orders of magnitude. The biological solution to the problem of visual perception is a highly structured process in contrast to smell. Vision uses many more processing states than smell, but its alphabet consists of only 32 shades of grey. We cannot discover experimentally how our brain works in visual perception, and, thus, we can only access it by means of computer simulations. However, experiments (Stryker, 1990) on the functional architecture of cat's and monkey's visual cortex demonstrate that neurons within radial columns share many specific response properties:

Specificity for topography, in that all neurons of a given column have their receptive fields in a particular portion of the visual field; it is structurally similar to the proximity law of human visual perception.

Specificity for orientation, in that all neurons of a given column will selectively respond to lines or edges; it is structurally similar to the equality law of human visual perception.

Specificity for cell types (on / off), which is related to brightness, some types firing at the same time; it is structurally similar to the closure law of human visual perception.

In any case, we might conclude that perceptual laws seem to be the result of the action of genes and of the emergence of new properties at the same time, while their linguistic manifestations look rather like the emergent property of a complex system, irrespective of whether they were finally fixed through evolution and incorporated by the genome, or were formed during ontogenesis every time a child acquires language. Sentences never group identical words together: *John came yesterday at home* is a sentence, **John John John John* or **came came came came*, are not. On the contrary, when a nose sniffs a scent, molecules carrying it are captured by the receptor neurons which, after being excited and after passing beyond a threshold, propagate the sensation until it reaches the olfactory bulb. And when our eyes look at an object and its image is reflected on the retina, there are millions of neurons that intervene to grasp specific features of that image: it is only afterwards that this dynamic gives rise to emergent attractors involved in a chaotic system.

We could hypothesize that these perceptual organizing devices, especially the visual ones, are the low level patterns that underlie the three linguistic and topological structures examined above. But there is a crucial difference between the two: merge (or confluence, its topological counterpart) joins two linguistic stimuli whose respective nature is asymmetric as only one of them projects its categorical features onto the entire chain; on the contrary, the closure law groups two stimuli together with neither of them prevailing in the resulting set. A similar difference exists with the asymmetrical joining of agree and pied-piping (ramification and beginning/end), on the one side, and of the equality law and the proximity law, on the other side.

Lorenzo and Longa (2003, 651), partially following Chomsky's assumption, consider that the set of computational operations (merge, agree and pied-piping) and, perhaps, a universal inventory of grammatical non referential features, are directly controlled by the genes, while other grammatical paths accomplished through derivation are the result of epigenetic processes. I can not entirely agree with them: in my opinion, these computational properties could (and should) also have arisen as a result of epigenesis, even so it is necessary to look for a more sophisticated explanation to understand the remaining ones, which are much more complex.

4.6 Epigenesis as a source for emergent properties

Epigenesis, an Aristotelian term opposed to preformationism, is one fundamental concept of current complexity sciences in order to explain the emergence of new levels in living beings, that is, the emergence of new wholes that are more complex than the sum of their parts. The epigenetic approach does not regard the information in the genome as a sufficient cause (although a necessary one) for the developmental process that makes up an organism. It has been characterized as follows (Solé and Goodwin) 61–62):

> One of the continuing enigmas in biology is how genes contribute to the process of embryonic development whereby a coherent, functional organism of specific type is produced. How are the developmental pathways stabilized and spatially organized to yield a sea urchin or a lily or a giraffe? The problem here is that genes are themselves participants in the developmental process. They do not occupy a privileged position in making decisions about alternative pathways of differentiation. Yet they clearly constrain the possibilities open to cells: lilies do not make muscle or nerve cells, giraffes do not make the water-conducting elements of plants. How do genes act and interact within the context of cells so as to bring about these units of structure and function? How do cells act and interact within the context of the organism to generate coherent wholes, the different types of organism that populate the planet? It is not genes that generate this coherence, for they can only function within the living cell, where their activities are highly sensitive to context. The answer has to lie in principles of dynamic organization that are still far from clear, but that involve emergent properties that resolve the extreme complexity of gene and cellular activities into robust patterns of coherent order. These are the principles of organization of the living state.

Thom pointed out that the potential pools catastrophe theory accounts for it belonging to the 'epigenetic landscape', as described by Waddington as early as 1940. Hence, it is interesting to consider the way new structures can emerge from a previous set of similar units in complex systems. S. Kauffman (1995) has developed an encouraging approach to the study of genetic dynamics in both development and evolution. The starting point of his model is that the activities of other genes determine the control of gene state, which varies along two positions: 0 and 1. Restricting the number of control signals that act on a single gene to two (connectivity $K = 2$), he explored the dynamics of genetic networks where all possible interactions among genes are allowed. The response of each gene to its two inputs was randomly determined in accordance with the set of sixteen Boolean functions of two variables. Following this pattern and applying it to a network of three related genes we obtain eight states separated into two trajectories each involving four states.

Formal inheritance 57

But when Kauffman increased the number of genes involved in the experiment, he found a surprising limitation of the dynamic patterns. On average, N genes resulted in $N^{1/2}$ attractors that cycle through a small number of states or settle at a stable point. If we interpret an attractor of the network as a differentiated cell type, then for the estimated 30,000 genes of human genome we obtain a picture that approaches the composition of human body.

The reason for this decrease of possibilities is *canalization*, which describes a property of Boolean functions relative to formal constraints. A canalizing Boolean function is one in which it is only one input which determines the entire output irrespective of the value of the other input. For instance, this would be the case in a X + Y > Z connection where we do not need to know the state of Y because whenever X is on (or off), Z is also on (or off). Putting it in a trajectory-like manner (in the sense of R. Thom), Boolean canalization would look like a confluence of X and Y where trajectory X continues until trajectory Z after the meeting point, whereas trajectory Y simply finished at this point.

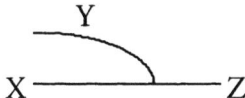

Merge, the most important principle of the computational system of grammar, has a formal structure that accommodates Boolean canalization quite well. The remaining principles, agree and pied-piping, are easily derived in a similar manner as they are also essentially asymmetric. This suggests how evolution could have implemented the perceptual system of a mammal with these new principles that produce human language. Perceptual patterns start from the undifferentiated background of similar units, be they neurons that perceive similar acoustic features, or neurons that perceive similar visual shapes; asymmetric patterns certainly arise when some acoustic or visual groupings are perceived as salient, but this step, which could be called the cognition of perceptual scenes, comes after. What drastically opposes language to any other representational system is the fact that it is based on asymmetry from the beginning. Mathematically we could characterize it as a system that takes account not only of a set of stimuli but of a previously organized one, of a topological space (López-Garcia, 1981).

The isomorphism that holds between the principles of the minimal program, on the one hand, and the perceptual gestaltic laws, on the other, is probably related to a characteristic of a minimal program that was emphasized by Chomsky: its optimality. We suppose MP differs from P&P because it establishes the conditions for an optimal linguistic system, for a perfect one. The word 'perfect' here means that the conditions of MP should satisfy the requirements of the interface levels, of phonetic and logical form. This is a consequence of a theory that is interested in the kind of knowledge a human language represents rather than in its adequacy for practical purposes. Linguistic utterances, Chomsky argues, are often ambiguous, and this is detrimental to the use of language. But linguistic utterances can harm neither the phonetic interpretation, nor the semantic one. It is recognized, however, that we do not exactly know how this adaptability to the interface levels should manifest. The above correlation suggests a way it could be done insofar as interface levels are perceptual levels that obey a set of laws which are isomorphic with those that organize syntactic sequences. This is the reason why in a cognitive approach to grammar the perceptual approach is clearly preferable to the cognitive: acoustic inputs are put together following the same rules that organize meanings into coherent groups, but only the latter may be said to constitute the object of cognition.

4.7 Protolanguage as a product of perceptual networks

The above hypothesis is supported by some independent proof. It is a well known fact that visual perception is the highest system of analyzing and storing information among superior animals irrespective of human language. If language, the most sophisticated informational system any animal species has ever developed, has to be explained in terms of evolution and natural selection, then the easiest first step is obviously the visual information system. In fact, both formal systems play a very similar role in the process of transporting the data of the external world into the organism. Both vision and language are re-presentational systems, that is, procedures that present pieces of contextualised information to the brain in a modified version again:

VISION: events > image of events in the retina > cognition of events
LANGUAGE: events > utterance relative to events > cognition of events

No wonder the respective brain paths of visual information and of linguistic information circulate quite close to each other. The traditional treatment of the issue does not support this claim because visual information is transported by the optic nerve from the retina to the occipital lobe whereas the centre of linguistic information treatment is located in the anterior zone of the brain, fundamentally around the temporal lobe. But Kaas (1989) has pointed out that further trajectories of the optic nerve surprisingly approach the speech areas: after having left the visual area, the optic nerve bifurcates into two branches, the dorsal that goes towards the parietal lobe and passes along Broca's area, and the ventral that goes towards the temporal lobe and passes by the Wernicke's area. However, as remarkable as it may be, this topological coincidence is not the most interesting fact. Kaas has also shown that each visual brain stream plays essentially the same coding role as its neighbouring linguistic area. Thus, the dorsal stream intervenes in the coding and decoding of visual moves and visual positions, but the ventral intervenes in the visual recognition of objects: correspondingly, Broca's area is the region in the brain involved in syntactic phenomena, but Wernicke's area is the region involved in semantics, that is, in selecting the lexical items that are referred to by the objects of the world.

Further evidence underlining the relationship between the emergence of language and vision is offered by Marr and his colleagues (1982) working on computer simulation of visual processing. They divided the process by which stimuli that impress themselves onto the retina finally produce a mental image in the brain into three steps:

i) A 2 dimensional scheme which consists of several uniformed regions that are separated from each others by borderline boundaries;
ii) A 2 1/2 dimensional scheme where the above patterns incorporate orientational frames by means of the introduction of specific vectors in all their regions.
iii) A 3 dimensional scheme, whose orientation changes from the object to the subject, that consists of the elaboration of a generalized image which is compared with a list of mental images stored in the brain.

These three successive steps could by paraphrased as follows: firstly, we distinguish the clear surface of the sheet of paper someone is grasping from the darker surface of their hand; secondly we incorporate the information that the sheet and the hand do not belong to the same plane, they form an angle; thirdly, we notice it is just a sheet of paper and a human hand.

It is interesting to compare these steps with the steps listeners follow in linguistic decoding: firstly, they hear a continuous sound sequence from the speaker and they proceed to split it in several parts with the help of some pausal breaks that punctuate the linguistic stream; secondly, each fragment receives a grammatical value (that is, a functional orientation) according to the morphemes that are attached to it; thirdly, lexical items are interpreted and receive a meaning that relates them to a referent in the world. Notice that these three steps successively performed by the listener correspond, though from the decoding perspective, to the three steps followed by the speaker in the minimal program: lexical items are selected and put together; a derivation is constructed by the computational system in order to provide these items with grammatical features that group them together; words are spelled out and interpreted by the interface levels, thus acquiring a phonetic and a logical form. The parallelism between this visual processing, as simulated by Marr, and the linguistic processing is obvious: it could be that evolution has converted the network of visual neurons whose regulating rules are genetically inherited into a similar set of nerve cells that work at linguistic processing and whose rules are inherited also in respect to this fundamental level. The only gain evolution has conceded the linguistic module is Boolean canalization, a device that permitted genetic networks to be improved with the asymmetric perspective that characterizes such phenomena as merge, agree and pied-piping require.

We mentioned the findings by Bickerton (1990) who demonstrated that the minimal syntactic structures of child speech before the age of two, the communicative skills psychologists taught to chimpanzees by means of Ameslan, and the grammatical patterns of pidgins fundamentally overlap. Bickerton called this step protolanguage, humans develop it every time they acquire a language, one can conclude that it is genetically imprinted on the human genome. This is certainly arguable, but for now, nothing will be said about it here. Any way Slobin (2002) has proved that Bickerton's proposal on Creole genesis revealing an innate "bioprogram" for language seems far less plausible than when it was introduced twenty years ago.

Much more interesting is the fact that the formal structure of protolanguage is very close to that of visual perception. For example, the following samples collected by Bickerton consist of groups of two or three items none of which prevail over the others either on formal or semantic grounds. These items share a common intonational pattern grouped together, they are attributed to the same semantic domain.

CHILD LANGUAGE	CHIMPANZEE LANGUAGE	PIDGIN OF HAWAI
Big train; Red book	*Drink red; Comb black*	*Ifu laik meiki*
Adam Checker; Mommy lunch	*Clothes Mrs. G.; You hat*	*mo beta make time*
Wall street; Go store	*Go in; Look out*	*mane no kaen hapai*
Adam put; Eve read	*Roger tickle; You drink*	*Aena tu macha churen*
Put book; Hit ball	*Tickle Washoe; Open blanket*	*samawl churen; house money pay*

This suggests that, no matter whether protolanguage is inherited or not, it is built up by the users of pidgin or by the children who begin speaking from the neural genetic basis that organizes visual perception. Every time humans are faced with a linguistic situation they cannot manage as usual, they benefit from the processing abilities of other perceptual systems.

When someone is unable to move because their legs are injured, they employ their arms grasping any rigid objects that are near them. Consequently, the human body is able to crawl, a move that functionally substitutes walk. Arms are not legs, of course, but they are better than other organs, like ears or noses, in relation to movement. This is also the case when we compare visual perception with verbalization. Somehow the mammalian processing of visual inputs can simulate the verbal processing by humans (that are mammals too). This is the reason why chimpanzees get some success in imitating human verbal behaviour, and it also explains how protolanguage can arise in pidgin situations and in baby talk. Perhaps, it even provides us with a plausible explanation for the early emergence of language. But language is much more that protolanguage. It is a complex system, a system that exhibits many formal features that no other biological system shares.

The importance played by vision in the origin of language does not exclude other senses from also playing a role. P. A. Brandt (2002) has argued that the pleasure music gives to people is extremely pervasive across cultures and it cannot be explained without supposing some kind of universal basis. He points out that there is evidence which proves that music and emotion are neurally interrelated through gesture and imaginary body schemas. Zatorre and others (2001) show there is additional bilateral activity when listening to music in occipital areas involved in visual processing, even without visual input. Hence, protolanguage was probably born from an integrated gestaltic layer which supports several senses, from auditory processing of music to visual imagery to motion. Nevertheless, senses form a hierarchy, touch, smell and taste being more primitive than hearing and vision, which only appear in superior animals. Protolanguage, the first step in language evolution, was surely more dependent on the latter ones than on the former.

5 How complex syntax can be?

However, Bickerton (2003, 87–88) is aware that protolanguage is not language as the latter has a very complex syntax:

> Perhaps the most depressing aspect of language evolution studies is fear of syntax, which, the present collection suggests [the Readings volume his contribution forms part of], is as widespread as ever. I know of no other field of study in which the work of a large body of highly intelligent specialists is so systematically misinterpreted, ignored, or even trashed. As a matter of plain fact, we have learned more about syntax in the last forty years than in the preceding 4,000, but you had never gues that from reading most books on language evolution, including, alas, this one. Syntax forms a crucial part, arguably the most crucial part – since no other species is capable of it – of human language. If we are going to explain how language evolved, we have to explain how syntax evolved. If we are going to explain how syntax evolved, we have to explain how it came to have the peculiar properties it has, and no others.

Although I do not share Bickerton's belief that most of our present knowledge on syntax was aquired in the last forty years alone, the centrality of syntax to the problem of the origin of language is certainly a point. Any proposal about the origins of language that fails to give an adequate explanation of its complexity will be mistaken. Linguistic complexity is two-rooted: there is a syntactical complexity, and there is a semiotic complexity. As this book is mainly concerned with the first one – although I will incidentally consider the second in chapter XI –, and the origin of language is a topic that is related to many epistemological domains, I will share a brief look at formal syntax with the reader.

5.1 On linguistic forms

Generative grammar made a very radical claim on linguistic structure about half a century ago: the properties of natural languages that the analysis of linguists had discovered are not worthless on their own, but they manifest a deeper form, the so called I-language (internal language). The goal of linguistics is little more than that of providing rigorous descriptions of the E-language (external language), but that of accounting for I-language. In other words: linguists should not be interested in what people speak about, but rather in what speaking people know. During the fifties the Chomskian turn looked very surprising to current linguistics, but it has become rather trivial at present time. However, generative grammar is far from deserving general acceptance among linguists. The reason is that viewing natural language as a form of knowledge may have two important implications:

First, language is (a form of) knowledge; Second, language is a form (of knowledge).

Nowadays hardly anyone may oppose the idea that language is a kind of knowledge. In fact, most linguists agree with it, although some of them also recognize that language is an action. It depends on the relative importance we adjust to the cognitive and the communicative facet of language that we are classified as a cognitivist or as a functionalist. But cognitivism does not coincide with generativism. Generative grammar also claims that the underlying principle of grammatical structure is not commensurate with the form of any other system in the world, or in the mind. This means that language has a specific form that is not reducible to psychological or social forms.

The challenge that generative grammar poses for Biology is due to this formal approach. No other theory of natural language has defended that Linguistics is a part of Psychology, which is a part of Biology, nor has it conceptualised language as the output of a biological organ, hard-wired in the human brain: Anderson and Lightfoot (2002) speak of the language organ, and claim that I-language is like the genotype that allows several phenotypes (E-languages). But Chomsky has had little to offer the evolutionary debate for he seriously doubts natural selection has played any role in shaping the structure of language. This scepticism stems from his belief that language is primarily a system of knowledge and, as a consequence, its form may not have been modelled by the selection pressures that operate in a social communicative system.

Within the generative paradigm the formal difficulty raised by Chomsky's proposal has never been completely solved. Therefore it is no wonder that Jenkins (2000) examined in great detail the three claims mentioned before, Goethe's, Saint Hilaire's, and D'Arcy Thompson's: he is aware of just how many difficulties the evolutionary aspect of grammatical form can represent if it has to support the generative paradigm. If we consider the classical approaches to this topic provided by the generative scholars, we find three positions (Knight, Studdert-Kennedy and Hurford, 2000): Lenneberg (1967) drew on a mass of data to see language as a self-contained biological system, but failed to derive gradually one step from the preceding one; Bickerton (1998), who showed how natural selection could have favoured increasing complex perceptual and representational systems in the brain, was not capable to account for the emergence of syntax in a similar gradual way, and he had, therefore, to admit a single crucial mutation that converted hominids in humans and rewired the brain in order to allow it to accommodate to the form of grammar; finally, Pinker and Bloom (1990) do not confront the problematic issue of form, concluding that language certainly benefited humans, and, as a consequence, evolved through adaptation because it provided fitness. Current approaches scarcely deviate from these ones, unless we introduce contextual considerations by means of ex-aptation or Baldwin effect. Unfortunately, no other alternatives remain: many linguistic forms seem to have emerged either through a series of successive catastrophic leaps, or as a gap that summed up a gradual evolution; otherwise, linguistic form simply vanishes, although linguistic form is precisely the problem we have to deal with. It is true that language has to be attributed not only to the evolutionary emergence of some kind of cognitive representational scheme in the brain, but also to the appearance of novel strategies of social cooperation (Deacon, 1997). But the linguistic forms that characterize the syntactic module of grammar does not reflect the structure of social action, excluding some well-known thematic patterns of verbs (a verb like *to give* needs an agent, a theme, and a patient, because the act of giving in real life consists of these roles, etc.).

5.2 Types of grammar: generative, cognitive, and functional grammars

How should grammatical forms be approached? According to Chomsky's proposals, there are two fundamental pictures, not necessarily unrelated ones:

A) The so called P&P (principles and parameters) theory claims that between the two empirical interface levels that any theory (and also any speaker) is forced to recognize, the PF (Phonetic Form) level and the LF (Logical Form) level, there is a specific syntactical framework that consists of other two levels, SS (surface structure) and DS (deep structure). Syntax would be a computational device that converts DS in SS under the conditions stated by the following modules: X-bar Theory, Q-Theory, Case Theory, Binding Theory and Control, Movement Theory.

B) The so called minimalist program (MP), which is aimed to improve P&P, proposes to simplify the precedent model in order to achieve explanatory adequacy. While P&P recognized, in addition to PF, three levels of syntactic representation, DS, SS, and LF, the Minimalist program, that tries to minimize the theoretical machinery of grammar, seeks to eliminate DS and SS, two levels that were set up on the basis of internal evidence, leaving only LF and PF. Since then, derivations are constructed by a CS (computational system) which selects items from the lexicon and arrives to a SD (structural description), which includes a pair of representations, LF and PF, that must satisfy the respective interface conditions. Syntactical principles are drastically cut off in MP: they can be reduced to Merge, Agree, and Pied-Piping.

Some readers may probably find that the two preceding paragraphs are hard to read or even obscure: In my opinion they may be right. I am not interested in carefully examining these models, neither P&P nor MP. In fact, many linguists do not work inside the generative grammar paradigm because they are convinced that most of the grammatical structures from languages in the world may be better explained by considering the contextual situation every structure verbalizes. Nonetheless, the formal structures generative grammar has repeatedly pointed out in their P&P version cannot be reduced to any perceptual or situational frame, and this poses a challenge to any theory of the emergence of language. Actually, when we speak of 'cognition and grammar', we mean two different things:

1) Grammars that are interested in the way linguistic structures reflect the patterns of the external world they refer to, and correspondingly in the patterns language imposes on the world: these are respectively cognitive grammars (Langacker, 1987–1991), and perceptual grammars (López-García, 1994–1998).
2) Grammars that are interested in what speakers know about language, and not in the meanings they are aware of employing the language: these are generative grammars of the P&P type.

Both are opposed to functional grammars, which study the use of language rather than its forms and structures. However, whereas cognitive and perceptual grammars study the knowledge of the world that linguistic forms manifest, generative grammars study the knowledge of the language itself. There is no reason for excluding any one of these three types of grammar because languages are used in speech acts, reflect an external world, and have a formal structure.

Any way, a fundamental question arises now: how should we account for the linguistic forms we are confronted with in everyday life? Functional forms, the forms that constitute the topic of functional grammars, are clearly a result of adaptation: evolution has carried off all linguistic patterns that did not favour the use of language. Cognitive and perceptual forms have also an adaptive origin: natural selection helped some mental and biological patterns to survive, and the linguistic forms that they reflected survived at the same time. Let us emphasize, however, that adaptation and natural selection do not affect any genetic endowment in the case of functional and/or cognitive patterns: the forms that natural languages incorporate as a result of contextual adaptation are always lenes (linguistic memes), never genes. In other words, they belong to the E-language, and not to the I-language: we may conceive a perceptual/cognitive grammar of Spanish or a functional grammar of English, but not a universal perceptual grammar or a universal functional grammar for perception/cognition and functions broadly varying worldwide.

On the contrary, abstract forms, those of generative grammar, lack any explanation relative to the external world: they simply are such and such. Does this imply that they are genetically encoded? As pointed out by Lightfoot (1999), some effects of UG (universal grammar) are dysfunctional. He compares them to patterns that evolve as a by-product of something else. And he suggests that a mental organ that manifests in language, in music, and in numbers, evolved with a capacity for discrete infinity with the particular properties we know based on the physical structure of the brain.

Unfortunately such an explanation does not account for the fact that language differs from numbers or from music in many respects. Ancient Greeks were already aware of the fact that music is related to Arithmetic, and the medieval syllabus of European universities included both in the Quadrivium. Language, however, belonged to the Trivium. It is possible that the form of I-language evolved as a by-product of something else, as claimed by Lightfoot, but it was not a mathematical structure. The hypothesis I am considering in this book is that the form of I-language corresponds to the form of the genetic code as such.

5.3 The formal syntax

As I said above, while the MP of generative grammar can be reduced to a very restricted set of formal properties which it shares with other formal or semiotic systems, this is not the case with the P&P paradigm. It is important to notice this in order to prevent a very common misunderstanding: the fact that Chomsky first proposed P&P and later proposed MP could be interpreted as if P&P were to be considered old and so not worth considering as a step of linguistic research. No wonder some scholars that work within the Chomskian paradigm are reluctant to accept this change in perspective (Newmeyer, 1998, 313–314):

> Chomsky's conception of the structure of UG [universal grammar] has changed considerably since the mid 1980s. The leading idea of the Government-Binding model of the 1980s [of P&P] was that the internal structure of the grammar is modular. That is, syntactic complexity results from the interaction of grammatical subsystems, each characterizable in terms of its own set of general principles. The central goal of syntactic theory was thus to identify such systems and characterize the degree to which they may vary (i. e. be 'parametrized') from language to language [...] In the Minimalist Program [the MP], however, the computational system consists of a set of unparameterized principles of a 'least-effort' sort that discard all but the most 'economical' derivations involving the same lexical resources [...] Now then, how does the move from the systems of principles of GB [P&P] to the economy principles of the MP impact the matters that concern us here? Assuming, following Chomsky, that grammatical principles are directly implemented in parsing, *GB [P&P] and the MP have opposite implications* [italics mine]. A plausible case can be made that many principles of GB [P&P] aid this process, in that their filtering function has the effect of reducing the set of possible candidate structures that the hearer of a sentence has to posit online [...] But the economy principles, as Chomsky is surely correct in pointing out, have precisely the opposite effect. Far from being local filters, they require, as we have seen, an immense amount of computation just to determine what derivations *might* be possible, followed by a cumbersome procedure of comparison of derivations and rejection of all but the one that is most economical.

In other words: P&P properties help the speaker/hearer to parse (to provide a syntactic structure for the sentence), and hence they are, in some senses, adaptive. Meanwhile, the MP model is unhelpful and, even, constitutes a drawback to the needs of current grammatical description, although it sheds a lot of light on the problem of the origin of language as we have seen. This is the reason why we can consider the P&P principles as a self-consistent set of linguistic features that any grammarian, not only a Chomskian, is forced to describe (Newmeyer, 1998, 315–316):

How complex syntax can be? 69

> The idea that UG is a by-product of the physical properties of big brains seems incompatible with what has always been taken to be one of its most characteristic properties – its autonomy. According to the autonomy thesis there exists a self contained 'grammar module', i.e. a system governed by a set of formal, discrete and interacting principles and rules. That is, underlying linguistic behaviour there is a separate component of our knowledge, the grammar, which is not reducible to other forms of knowledge. But we have a contradiction here. UG cannot be derivative and autonomous at one and the same time. What are the chances of UG emerging as an automatic consequence of any set of external principles, but having an internal algebra totally independent of these principles? Absolutely none, I would venture to say.

I agree. This does not mean that the MP proposal is worthless, but it is much more closely related to the computational properties of the hard-wired brain than to language. Universal grammar, whatever this word may mean, is not the consequence of the efficient packaging of neurons in a big brain, it is what makes humans unique, and, thus, a set of arbitrary, mostly dysfunctional (Martin and Uriagereka, 2000) principles that every human language possesses and that are apparent to each human speaker/hearer. These principles, though collected by Chomsky in the P&P model, are neither specifically Chomskian, nor generative. In fact, the formal laws we will consider next are usually recognized and have been listed by grammarians since structural linguistics was born in the 1930s, in Europe with the Prague and Copenhagen schools, and in America with I.C. Analysis and distributionalism.

5.4 A glance at the syntactic principles

Consider the following advertisement (as this paragraph is only intended to help those readers who are not grammarians, linguists may pass it by):

> *We are delighted you have decided to buy our portable computer and hope you will enjoy it and benefit from using it at home, on holiday, or at work*

5.4.1 Dependence relations characterize the summing up of lexical items

Suppose the reader knows the meaning of all the words that constitute the complex sentence above. This would be not enough to understand it. The reader / listener is able to draw more information by means of parsing, that is by adding a structural description, which establishes a hierarchical web of dependencies, to the sentence:

1: add *portable + computer*
2: add *our + portable computer*
3: add *to buy + our portable computer*
4: add *have decided + to buy our portable computer*
5: add *you + have decided to buy our portable computer*
6: add *are delighted + you have decided to buy our portable computer*
7: add *we + are delighted you have decided to buy our portable computer*

Grammarians have also proposed other alternative analyses I will not examine here: in fact, most scientific research into grammar consists of contrasting several proposals to analyze the same text. The point, however, is that they always agree on those structural analyses that must be rejected, for example:

*add [*we are*] + [*delighted you*] + [*have decided to*] + [*buy our portable*] + [*computer*]

The reason is that the lexical information, which is stored in the mind, is syntactically represented, thus an unacceptable ordering of the added lexical stretches is excluded. Generative grammarians call this property the *projection principle* (Chomsky, 1981, 39), which establishes that the thematic (actantial) structure associated with the lexical items never changes in the course of a derivation and must be saturated in the syntax. Structural grammarians call this *valence*, and conclude that every predicate has one or several valence webs which determine the number and characteristics of the arguments that are related to it (Tesnière, 1959; Helbig, 1982).

5.4.2 Dependence relations are established among heterogeneous units

The complex sentence above is like a building that has been obtained by heaping up some previously made bricks. However, this kind of analogical comparison can go no further. Notice that in a building the bricks at the top of the wall must be layed after the lower ones, but masons pick them up from a pile of undifferentiated bricks. This is not the case in syntax, we do not perceive the sentence as if a set of words had successively been joined together. We do not add *portable* and *computer* as if they were two separate entities (i.e., this is a *portable*-brick and this is a *computer*-brick), but *portable* (or *fixed*) is a property any *computer* includes on its own. Consequently, we may or may not add *portable*, but *computer* is obligatory (hence, we can say *to buy our portable computer* or *to buy our computer*).

Generative grammarians call this property the *X-bar theory*, which brings out what is common in the structure of phrases. According to the X-bar theory, all phrases are headed by one head which is considered to be a zero projection (X^0); heads are terminal nodes, the dominant words which reflect the main perception of the world each phrase is aimed to represent. Moreover X-bar theory distinguishes two further levels of projection: complements combine with X to form X'-projections; the specifier combines with the topmost X' to form the maximal projection XP (this is the classical proposal by Jackendoff, 1977: lately other scholars consider that the specifier is the head). For example, our portable computer would be analyzed as:

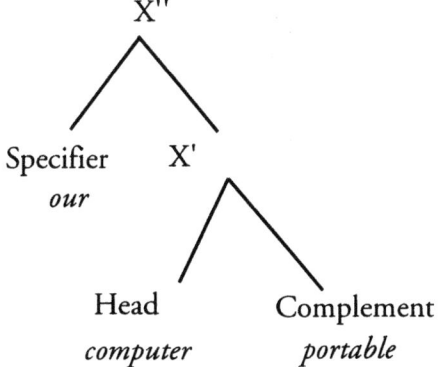

Functional grammarians call this a *determination relationship* (Bally, 1932, chap. 3; Trubetzkoy, 1939): every time two words are put together, one of them, the *déterminant*, is said to complete the other, which is the most important one, the *déterminé*; as a consequence of this, a new situation, the *défini*, is created.

5.4.3 Agreement: some overt relationships help to solve adjustment problems

Verbal worlds and mental worlds cannot remain mutually unrelated because verbal occurrences are employed for the expression of mental processes. However, the first are linear and language specific, the second are not, and both worlds can diverge in several degrees: every human language is used to manifesting the mental process which underlies a sentence by means of a specific set of features, some of which reflect the structure of cognition more directly than others. For example, consider the verb *to eat*: the range of words that can function as its direct object is very limited, they are necessarily food or some kind of edible entities. On the contrary, the subject of *to eat* can be any animate being or, even, any concept metaphorically relatable to it (as in *inflation ate up their savings*). The direct object is, hence, an internal argument of the verb which closely reflects the structure of perceived reality, whereas the subject (if there is a subject in the language) receives an extra feature that ties it strongly to the verb: "agreement". Thus, in the sentence above *you will enjoy it and benefit from using it* the subject *you* agrees with the verb *benefit* but the object *it* does not (as a subject *it* would agree in the way *it benefits*).

Agreement relates the inflectional properties of the verb to the subject, which is forced to repeat them. Generative grammarians (Gazdar, Klein, and Pullum, 1983) recognize a node INFL which is taken to contain all verbal inflections, i.e. including person and number properties. As the so called *Case theory* assigns an abstract case to every overt noun phrase, INFL assigns a nominative case under government to the subject, which is the origin of agreement. The agreement relationship in some structures that clearly display it has been recognized by classic grammar since Greek and Latin scholars began to study the morphology of their native languages. Functional grammarians inherited this tradition describing agreement by means of the concept of "selection"; for example, Bloomfield (1933, 12.6) characterizes *congruence* as the simplest kind of agreement which is defined as follows: if the actor is a form of sub-class A, the action must be a form of sub-class A.

5.4.4 Anaphora and ellipsis as cohesive procedures

To speak a language is to re-present (to present again) the perceived world by selecting only some of its characteristics. When we say (*this is*) *a good table*, we are ignoring the fact that the table is m*ade out of wood*, that it is *round*, that it is *brown*, etc. When we say *look at the book*, we are ignoring the fact that the book was one *about Chemistry*, but we are also ignoring our wish that the listener looks at it *now* or *in the office*, etc. This means that the linguistic utterances always leave many things unexpressed, some of which have to be recovered by the listener if he/she wants to be able to understand the utterance.

The listener recovers the elements of the context that are lacking, sometimes directly – this is ellipsis –, and sometimes because the speaker points at them with an index – this is anaphora –. For instance in *and hope you will enjoy it*, there is an elliptical subject *we* (*[we] hope you will enjoy it*) and there is a pronoun *it* which anaphorically substitutes *our portable computer*: the listener would never be able to understand the meaning of the advertisement above without knowing how ellipsis and anaphora (pronouns) work.

Generative grammarians describe ellipsis and anaphora by means of two specific modules (Heny, 1981). The module of grammar regulating NP interpretation is referred to as the *binding theory*; it is responsible for assigning an appropriate interpretation to the idealized NPs in sentences like the following one: *John$_m$ admires her$_k$*; *Mary$_k$ hurt herself$_k$*, where m, k are referential indexes. Three types of NPs are distinguished: full noun phrases, like *John*; pronouns, like *her*; reflexive elements, like *herself*: reflexives are bound to a NP in an argument position (they are /+anaphor/); pronouns should not be linked to a NP in an argument position within the binding domain (they are /+pronominal/); referential expressions must not be bound by NPs in argument positions (they are /-anaphora, – pronominal/). The module of grammar regulating non overt elliptical categories is referred to as the *control theory*: PRO, which occurs as the subject of non finite clauses, is a non overt NP, a NP which is syntactically active, but which has no overt manifestation (it is /+anaphor, +pronominal/). According to control theory, PRO must be ungoverned, and it is controlled by another NP in the sentence: in *you have decided* PRO *to buy*, the subject PRO (i.e. *you*) of *to buy* is controlled by the subject *you* of *have decided*.

However abstract and technical the above description may look, it is interesting that current explanations of ellipsis and anaphora by functional grammarians do not substantially differ from it. Although functionalists (Halliday and Hasan, 1976) recognize almost the same set of properties, – they consider these properties appear not only at sentence and phrase levels, but also at a higher non structural level: the text. This gives their papers a less formal and more descriptive appearance than generative papers on the same topic. Functionalists think texture (i.e. the property of being a text) is a *cohesive relation* that links the parts of a text together without being a structure. Texture consists of several cohesive procedures, fundamentally of five types: reference, substitution, ellipsis, conjunction and lexical cohesion, the former (reference and substitution) related to binding theory, the later (ellipsis and conjunction) to control theory.

5.4.5 Categories

Categories such as noun, verb or adjective have belonged to grammar since the Greek authors became interested in their own language. Nevertheless, until modern linguistics was born in the 20th century, categories were considered to be the expression of ontological values: the noun would be the linguistic procedure that permits us to manifest things, the adjective would be the linguistic procedure for qualities, and so on. It is obvious that such definitions, although philosophically interesting, are worthless in linguistics for certain words belonging to a category in language A and to another category in a language B. This is the reason modern linguistics approaches the category problem from a formal point of view: according to Harris (1951, chap. 15), a category is a formal class of morpheme, that is, a series of morphemes which substitute for each other in one or several frames.

In generative grammar categories are dealt with following this pattern, but emphasizes the upper classes, rather than the sub-classes as Harris and other distributionalists did. Distributionalists divided the general class verb, which was established in accordance with a selected number of environments their members have in common, and afterwards they derived several sub-classes, which are groupings of morphemes in respect to all occurrences. Generative grammarians, on the contrary, relate the category verb to the category noun, for instance, because it is possible to pass from a verbal phrase to a noun phrase through transformation (*John discussed the project* > *John's discussion of the project*).

5.4.6 Movement relates different orderings of the chain of words

In terms of the history of linguistics, transformations are not an original contribution of generative grammar, but of structuralism, for they immediately follow the problem of categories, although it is true that generativists developed this line (as shown by the papers collected in Jacobs and Rosenbaum, 1970). Harris (1951) established a set of procedures that enables classes of morphemes to be obtained, in such a way that each morpheme can be substituted for the other morphemes of its class in an utterance. Next, new procedures are established which equate sequences of classes of morphemes. Finally, the relation between a morpheme class in one position and the same class in other positions gives cause for introducing the concept of transformation.

Harris was only interested at sentence level and lower, but researchers of the Prague school came to very similar conceptions working at text level. According to them (Firbas, 1964; Daneš, 1967), words are not arranged just any how: although each language has its preferred patterns of order, it is a universal law that items that rely on known information come first (the theme), and those that depend on them come after (the rheme). For each utterance arises as a response to the environment, some variation of order is always allowed: *Mary will come tomorrow*, whose subject *Mary* is the theme and whose predicate *will come tomorrow* is the theme, could be a response to an environment where the listener is looking for Mary; on the contrary, *will Mary come tomorrow?* would be a response to an environment where the listener hesitates about the behaviour of Mary (the auxiliary *will* being, thus, the theme). Order variation requires that certain morphemes may be "moved" in order to relate two different positions of the same morpheme: *Mary will come tomorrow* -INT-> *will Mary come tomorrow?*

The treatment of order and movement in generative grammar distinguishes two levels of syntactic representation: *D-structure*, a level which encodes the lexical properties of the constituents of the sentence, and *S-structure*, a level which reflects the more superficial properties of the sentence (the actual ordering of the elements in the surface string). These very two levels are also found in Tesnière (1959) when he opposes *l'ordre structurale* to *l'ordre linéaire*. Movement, in its turn, is explained in generative grammar as follows: A'-movement moves an argumental constituent to a non argumental A'-position leaving a coindexed trace in the bare position. For example, we obtain *will Mary come tomorrow?* starting from *Mary will come tomorrow* and putting *Mary* behind the auxiliary verb *will*, which is a non-argumental position, whereas a trace is left in subject argumental position: [...] subject *will Mary come tomorrow?* Either way, the treatment of movement is still problematic in the current paradigm of generative grammar: while the earliness principle establishes move as soon as possible, the procrastinate principle will delay it as long as possible.

5.4.7 Restrictions of movement

These disagreements would not matter if movement were an unrestricted operation. But movement cannot be unrestricted. If it were, some morphemes would pass the boundary between sentences, and disturbances in meaning would immediately appear. For instance, we cannot move *at work*, in the advertisement above as far as the first clause passing it through the boundary because this would change the meaning of the text: *at work we are delighted you have decided to buy our portable computer* means that we are delighted *at work*, but not *at home*.

In order to restrict movement, generative grammarians postulated the *subjacency condition* (Chomsky, 1981): movement cannot cross more than one bounding node. Functional grammarians faced up to these very restrictions by means of the concept of *isotopy* (Greimas, 1966): every sentence in a text belongs to a given isotopy (~story) and it has to remain inside; this permits the sentence to benefit from the meanings that previously appeared in the story without it being necessary to mention them again every time.

5.4.8 The construction of texts in speech acts: focus

But the maintenance of an isotopy is not only guaranteed by restrictions on movement. In real language, that is, in speech acts, meaning is cooperatively constructed by the speaker and by the listener: hence turns that follow in the speech stream also have to be coherent. In this case, new types of formal units appear which rely on the information of the preceding turn in a kind of congruence between successive turns: focus. When someone says *it was the 4,30 train that was delayed*, we may infer that in the previous turn the speaker alluded to another train (*the 7,20 train was delayed*) or simply asked about the trains that were delayed (*which train was delayed?*). Similarly a high pitch on a word as in *the train was delayed for TWO hours* implies that in the preceding turn someone made an opposite claim such as *the train was delayed for one hour*. The treatment of focus and other related topics was first undertaken by the Prague school of functional sentence perspective (Daneš, 1964) and later reelaborated by generative grammarians (Zubizarreta, 1998).

5.5 On the significance of the formal principles of syntax

The above principles are universal, no human language being conceivable without them. Generative grammar joined them together in the P&P model, but functionalist approaches also reached them, as I have shown, despite the different names each principle receives in several functionalist or structuralist schools, or, even, no name at all.

Readers may be surprised by the fact that generative grammar has reached a unified inventory of these principles, whereas functional grammars have not, it being necessary to collect them together from several theoretical proposals. And what is worse, they could think that this book is intended to justify the formal principles of generative grammar – by showing their genetic background –, and not the reason why those principles are somewhat irrespective to the linguistic school that discovered them. The difference between generative and functionalist approaches lies in the different properties each believe in: generative grammar is interested in the formal properties of syntax that lack an explanation in terms of communicative or cognitive adaptation; functional grammar, on the contrary, is primarily interested in those features that reflect the properties of human cognitive or communicative behaviour. Consequently, the set of formal principles above looks rather sparse and unjustified in functionalist papers, although, conversely, the way they approach the study of text – a non structural part of language – is much deeper than the generative proposals on the topic.

Generativists would probably say that the current paradigm of generative grammar, which is MP, does not challenge functionalism, and that there is no sense in starting from P&P, an old-fashioned proposal. But it is important to emphasize that MP and P&P are not new and old hypotheses, but two independent proposals. As Baker (1997, 129) says:

> Linguistic representations and conceptual representations are two different things [...] Subject and object are syntactic notions, defined by the language faculty, while agent and theme are conceptual notions, defined as conceptual representations [...] Chomsky's Minimalist conjecture [says] that language is in some sense an optimal way of satisfying "bare output conditions" defined by the language-external systems. However, it is worth observing that this minimalist conception significantly blurs the distinction between P and P theory and functionalist approaches to language, which characteristically de-emphasize syntax as a separate study and focus on its connection with cognition, lexical semantics, and discourse pragmatics.

Baker is right: the Minimalist program of generative grammar represents a kind of functional justification of the origin of language, even though it seeks the explanation in the structure of the human brain rather than in culture. I have tried to justify the Minimalist program above as an investigation into the fundamentals of human protolanguage: in this sense, it represents an optimal communicative solution for it allowed hominids to develop a protolanguage.

But between the step represented by protolanguage and the step represented by current languages, which are a culturally and cognitively determined diversification of universal grammar, there is a step that we are, as yet, unable to understand: the stage of the formal properties of syntax. Those steps in evolution, like protolanguage or current languages that exhibit functional properties, do not rise any epistemological problems, though some details may remain obscure or unknown: evolution is a matter of adaptation, and adaptation is functionally dependent. But the fact that all human languages exhibit the same set of formal syntactic properties, and that they are not functional, hence not adaptive, seriously challenges the picture we are establishing. How did they appear? Why are they precisely these properties and not others. The following chapters aim to answer these intriguing questions.

Before I go on, it is important to consider what I am trying to say when I characterize these formal principles of syntax as neither functional nor adaptive. It is easy to imagine many other formal systems that would fit the formal requirements a combinatory procedure for human language is obliged to satisfy. In fact, Chomsky (1959) observed that the syntax of language is simply one of hundreds of semi-Thue Boolean algebras, and computers make use of formal combinatorial systems that permit them to parse (to give structure to) information chunks in a very efficient and non human way (Gazdar and others, 1985). This means that the formal properties of syntax above (§5.4) are dysfunctional because they do not reflect any obvious communicative necessity, but this does not mean that they are worthless. In other words: the formal syntax of human language could have been different, but without formal syntax the communicative instrument would have collapsed.

Actually, the formal properties I have considered in § 5.4 are mutually dependent:

the world is perceived by means of signs
↓
signs are organized in lexical categories
↓
items that belong to lexical categories determine the kind of lexical categories which may combine with them
↓
these combinations distinguish a dominant head and several dependent components
↓
some combinations are covert and some are also overt: covert-overt combinations reinforce the link by means of a formal device (agreement)
↓
some other structures are mainly overt: anaphora and ellipsis tie up the sentences of a text
↓
movement transformations altere the appearance of the text
↓
movement is, however, a limited possibility for it helps to maintain the unity of the text
↓
the text relies on the context and this causes new movement possibilities

The formal properties grammarians are used to recognizing in the syntax of languages are, then, individually dysfunctional, but globally coherent. This means that they constitute a framework. Human language could have been otherwise, but language cannot develop its cognitive and/or communicative functions without joining words together in a text. And the instrument that permits language to do so is precisely this formal framework.

6 A blind alley

6.1 A striking parallelism

The state of art has changed a lot since the *Société de Linguistique de Paris* prohibited any kind of research on the subject "the origin of language" in 1866. For the last two decades, linguists have been increasingly interested in the forbidden topic. By adopting a weaker version of Darwin's theory, they are trying to prove that language can be biologically rooted, and explained by natural selection, just as Darwin himself proposed. However, for the time being, these attempts still remain very tentative, and neither linguists nor biologists seem to take them seriously. Perhaps this is due to the fact that they have tried to describe the evolution of language before demonstrating its fundamental biological nature. If language were not a biological entity, but a cultural one, there would be no need to investigate how natural selection could act at the beginning of language evolution.

This does not mean that linguistic considerations have been excluded from research conducted by geneticists. On the contrary, the understanding of the genetic code is used to employing a linguistic metaphor that compares the nitrogenous bases (the base-sugar-phosphate groups being nucleotides) with the letters of an alphabet, the codons with the spelling of words, and the amino acids with their meanings. By doing so, geneticists did not aim to suggest any functional or evolutionary resemblance between the genetic code and the linguistic code. They simply benefited from an elegant and apparently suitable analogy for pedagogical reasons.

This striking parallelism lies implicitly at the basis of the following statement made by M. Ridley (1983, 10–11):

> The outstanding example of a universal homology is the genetic code [...] What matters here is that the code, although it is arbitrary, is known to be universal. *It is arbitrary in the same sense that human language is arbitrary* [italics mine]: there is nothing about a horse that demands it must be specified by that sequence of five letters; any sequence of letters would do. The genetic code has the same property [...] The universality of the code is easy to understand if every species is descended from a common ancestor. Whatever code was used by the common ancestor would, through evolution, be retained. It would be retained because any change in it would be disastrous. A single change would

cause all the proteins of the body, perfected over millions of years, to be built wrongly; no such body could live. It would be like trying to communicate, but having swapped letters around in words; if you change every 'a' for an 'x', for example, and tried talking people, they would not make much sense of it.

Ridley's comparison constitutes a very frequent resource of the popular paperback introductions to Genetics and even of university guide books: genetic code is like a linguistic code. Then, the sentence I have underlined above is by no means noticeable. But the context in which it appears could have given rise to an innovative expansion of this homology, which would perhaps have read as follows: *genetic code is universal in the same sense as linguistic code is universal* [italics and text mine].

Ridley does not extend the parallelism in this direction and the reason is that organisms would collapse when we substitute one letter for another, but texts would not. There are many ways of writing a message in code, but the substitution of one type of letter for another – say, every 'a' by every 'x' – only would give rise to a rather naïve and easy to be decoded new text. Even if we substitute one natural language with another, comprehension is still possible by translating the text encoded in the superficial code A (f. ex. Spanish) into a text encoded in the superficial code B (f. ex. English). The set of linguistic conventions that must remain unaltered in any case are those of the deep code, just the grammatical code all human languages share in common. The alteration of such a code would produce a severe jar on communication between the members of two successive generations. It would be as if parents had spoken a natural language and infants had mastered an artificial one, for instance, a computer code based language. Such an informative codequake never took place: if it had done, mankind would have converted into a very different species. And the conclusion we are tempted to draw is the following: *the reason why the grammatical code has never changed is because it is formally determined by the genetic code, which has also remained unchanged since it appeared on earth*. This is a very hazardous hypothesis, of course. I will therefore devote the following chapters, in an effort to compound it.

Thirty years ago, the well known linguist R. Jakobson (1971, 676–682) pointed out that this metaphor perhaps hides much more than we suspect:

> Human language is, as biologists term it, species-specified. There are in any infant innate dispositions, propensities to learn the language of his environment [...] The spectacular discoveries of the last few years in molecular biology are presented by the explorers themselves in terms borrowed from linguistics and communication theory [...] From the newest reports on the gradual breakthrough of the DNA code [...] we

actually learn that all the detailed and specified genetic information is contained in molecular coded messages, namely in their linear sequences of 'code words' or 'codons'. Each word comprises three coding subunits termed 'nucleotide bases' or 'letters' of the code 'alphabet'. This alphabet consists of four differing letters 'used to spell out the genetic message'. The 'dictionary' of the genetic code encompasses 64 distinct words which, in regard to their components, are defined as 'triplets', for each of them forms a sequence of three letters. [...] The transition from lexical to syntactical units of different grades is paralleled by the ascent from codons to 'cistrons' and 'operons', and the latter two ranks of genetic sequences have been equated by biologists with ascending syntactic constructions, and the constraints on the distribution of codons within such constructions have been called 'the syntax of the DNA chain' [...] The strict 'colinearity' of the time sequence in the encoding and decoding operations characterizes both the verbal language and the basic phenomenon of molecular genetics, the translation of the nucleic message into the 'peptidic language'. Here again we come across a quite natural penetration of a linguistic concept and term into the research of biologists who, by collating the original messages with their peptidic translation, detect the 'synonymous codons' [...] How should one interpret all these salient isomorphic features between the genetic code which 'appears to be essentially the same in all organisms' and the architectonic model underlying the verbal codes of all human languages and, *nota bene*, shared by no semiotic systems other than natural language or its substitutes? [...] Now, since 'heredity, itself, is fundamentally a form of communication', and since the universal architectonic design of the verbal code is undoubtedly a molecular endowment of every Homo sapiens, *one could venture the legitimate question whether the isomorphism exhibited by these two different codes, genetic and verbal, results from a mere convergence induced by similar needs, or perhaps the foundations of the overt linguistic patterns superimposed upon molecular communication have been modelled directly upon its structural principles* [italics mine].

For the time being, this question has not yet been answered. I do not think this is because it was inappropriately raised. As Jakobson emphasizes, no other semiotic system is governed by the rules that the genetic code and linguistic code share in common, and this fundamental resemblance ought to have a deep significance. One can imagine many other structures for the genetic code, and many others also for the linguistic one. The former appeared on earth by chance: the hypothesis we would like to challenge is that the latter was also randomly born. There are other codes that slightly differ from it in their effects on communication. For example, computer simulations of verbal behaviour are not far from linguistic interactions, although they are based on a very different type of code. One can perfectly conceive of a language whose formal code is like that of a computer. But, as a matter of fact, we know it is not, and this is the fundamental fact we are compelled to begin with.

6.2 Some inadequacies of the comparison

Unfortunately, however, the starting point of the biologists, which Jakobson reported on, is misplaced. The first inaccuracy of the comparison lies in the supposition that codons are a kind of sign in the same sense as words are. Signs are structural units that consist of two unrelated levels. In natural language these levels are the sound level and the meaning level (the *signifiant* and the *signifié*, as De Saussure called them). The word *cold* is the sum of the sound ['kould] plus the meaning cluster "chilly/lacking warmth of emotion/a low subnormal temperature". Both levels have nothing in common: sounds are physical entities and constitute the object of study of Acoustics, a branch of Physics; meanings are cognitive entities and they do not exist beyond social life, they should be the object of study of Human Sciences like Psychology or Sociology. No correspondence can then be established between sounds and meanings, and for this reason De Saussure claimed that the linguistic sign is arbitrary.

On the contrary, the bases that form part of the codon and the amino acids that constitute their reading are entities of a very similar chemical nature: the former are acids with Phosphorus (-POOH), the latter are also acids, but with Carbon (-COOH); both have Nitrogen:

$$
\begin{array}{cc}
\begin{array}{c} NH_2 \quad OH \\ | \quad\quad | \\ H - C - C = O \\ | \\ R \text{ (base)} \end{array}
&
\begin{array}{c} OH \\ | \\ O = P - R(\text{pentose}) - R' \text{ (nitrogen base)} \end{array}
\\
\text{AMINO ACID} & \text{NUCLEIC ACID}
\end{array}
$$

Whereas linguistic translation is between two domains that have no elements in common, genetic translation establishes a relationship between two domains that are partially identical. As a result of this fundamental difference, the linguistic metaphor, as employed by geneticists, never had any explanatory power. It was undoubtedly useful as a descriptive tool, but it never inferred consequences by examining the linguistic structure of the genetic code.

This first step in the wrong direction has been responsible for a cascade of untenable pairings. The next step, of course, was to consider that bases are like letters (or sounds), for a word consists of a string of phonemes from the phonological point of view. This results in the formula "genetic code has four letters (A, T, C, and G) that can combine together to form 64 (= 4^3) words of three letters each". And the habitual expositions of the metaphor observe that since only 20 amino acids are used in making proteins, neither codons consisting of one single base (4^1 = 4) nor consisting of two (4^2 = 16) would have been enough to do the job of giving instructions for 20 amino acids. This is true, but one wonders why we do not have codons made out of four bases (4^4 = 256), of five bases (4^5 = 1024), and even more. Natural languages make use of twenty phonemes on average, and words are strings of phonemes that have no more than about fifteen phonemes each, but the resulting number of words is much less than $1^{20} + 2^{20} + 3^{20} + 4^{20} + 5^{20} + 6^{20} + 7^{20} + 8^{20} + 9^{20} + 10^{20} + 11^{20} + 12^{20} + 13^{20} + 14^{20} + 15^{20}$.

The selective advantage of the triplets of bases is supposed to be that they prevent the code from mixing codons when a mutation takes place. For example, the amino acid Alanine is represented by four codons, GCA, GCC, GCG, GCU: if there were only 20 codons, each one corresponding to one amino acid, every time a mutation appeared in the third position the codon would be confused with another codon; on the contrary, since there are four synonyme codons, the resulting amino acid continues to be Alanine no matter which mutation occurs in that position. But such an advantage would be attained much more easily with 256 codons of four letters: there are a lot of complex and energetically expensive repair procedures in the organic world that prevent the genetic code of the confusions caused by mutations. They could dispense with. Natural languages, and actually, utilize much more the elongation of the string of phonemes that constitutes the word than any other repair procedure such as repetition.

Moreover, the metaphor we are considering here is not only inadequate in terms of quantitative arguments. It also fails on qualitative grounds. In the DNA strings that constitute the genetic endowment of an organism it is sufficient for a single base to be changed, deleted or added for the entire peptidic reading getting disturbed. For example, if we consider the string:

$$\ldots \text{GCA GAA UUA AGA GUA} \ldots$$
$$\text{Ala Glu Leu Arg Val}$$

and the second G is suddenly deleted, then we obtain:

$$\ldots \text{GCA AAU UAA /}$$
$$\text{Ala Asn stop}$$

where a stop instruction introduced by UAA has the effect of interrupting the reading and presumably of causing severe disturbances. This is not the case of natural languages. Single phonemes or letters of a text are constantly changed, deleted or added without communication getting altered. Sometimes changes belong even to dialectal variation, as when *do not* is simplified as *don't*, but in any case the replacement of any one letter rarely causes disturbances. This is due to the fact that the context helps to maintain the intended meaning, and especially because the deletion or the addition of a letter never alters the pattern of reading or produces the breakdown of the boundaries of the words. For instance, the substitution of the phoneme /ch/ of *march* by the phoneme /sh/ produces a different word, *marsh* (actually, the method of Phonology proceeds in this way); however, in a sentence like *they arrived in march before the beginning of the fair* no one is tempted to interpret the word *marsh* of *they arrived in marsh before the beginning of the fair* as a kind of swamp. Much more important is the fact that any changes in the letters (phonemes) of a verbal text always leave the boundaries of its words unaltered: for example when the second *s* of *she has the nice one* is lost, that is, *she ha[…] the nice one*, we do not obtain *she hat hen ice one*. This is a fundamental difference between linguistic code and genetic code which thwarts any attempt to compare them on these grounds.

There are other possibilities, of course. One of them would be to consider bases as morphemes and amino acids as the meanings of words. For instance, *cashier* would be analyzed as [cash] + [-ier] = "officer of a bank", *untimely* as [un-] + [time] + [-ly] = "inopportune", and so on. This interpretation mirrors the genetic code much more closely than the preceding one. A language like English has several hundred morphemes and they are gathered together to form several thousands of words. The magic proportion of 4:20 of the genetic code, that is, the existence of five amino acids for each nucleic base (1:5) is not far from this. Besides this metaphor satisfies the structural condition that was absent in the following one: morphemes and words are not identical entities but they certainly belong to very close linguistic classes, both consist of sound-meaning units. The parallelism with the genetic code would read this way:

GUA   part- -(n)er- -ship

"Valine"   "association"

However, the proposed interpretation has the fundamental drawback that morphemes and words that are made out of them are not only similar on formal grounds, as required, they are also alike from a semantic point of view. As a matter of fact, words *consist of* morphemes: the word *partnership* is the sum of *part* + the affix *-(n)er-* that means the person + the affix *-ship* that means a quality shared by the members of a group. With this in mind, any speaker of English that has never heard the word *partnership* could deduce its meaning by simply knowing the meaning of *part*, then comparing *partner* with *sleeper, worker*, etc., and finally with *friendship* and so on. Word formation by means of morphemes is not arbitrary, the morphemes that are joined together in a word contribute to its further understanding by adding their own meaning to it. This process reminds us of combinations in Chemistry: NaCl is a salt made out of Na^+ and Cl^-, and its chemical behaviour is a consequence of the chemical properties that characterize their components. But biochemical compounds do not behave this way: what distinguishes them from the former two examples is precisely the fact that Guanine + Uracil + Adenine give rise to the amino acid Valine, however, no adduced chemical property is able to explain the cause of this. The association of a codon consisting of three nucleic bases to an amino acid is arbitrary, apart from the topological conformation of the tRNA that hooks one onto the other; the association of the elements that form a chemical compound is not arbitrary. This is why our proposed comparison is so unsuccessful even in the present case.

We could try to compare the formal aspect of the biochemical equation to some kind of syntactic items, for instance to the syntactic functions or to the thematic roles. This was the proposal I made in López-García (1998) and, although it satisfies most of the prerequisites, it still lacks some conditions for it to be entirely adequate. Suppose we compare the codon to the string of syntactic functions within a clause, for example, the subject, the predicative functor, and the object:

subject + predicate + object

"the child   ate   an apple"

As one can see, the notions 'subject of', "predicate' and 'object of' are of an entirely formal nature, they are grammatical meanings, not lexical meanings. This allows them to fulfil the arbitrariness condition: the formal level and the semantic level are joined by chance. On the other hand, these formal functions cannot be deleted, changed or added without immediately affecting the semantic level: for instance, the deletion of the first function yields an ungrammatical clause, *ate an apple; the addition of another object yields an ungrammatical clause as well, because "subject + predicate + first object + second object" accepts verbs like *to give* (*the child gave an apple to her brother*), but not the verb *to eat* (**the child ate an apple to her brother*). In this sense they could be considered as suitable candidates for the genetic metaphor.

Nevertheless, the consideration of syntactic functions as the linguistic counterpart of genetic bases also has some disadvantages I was not aware of when I first proposed them. The problem is two sided. First, syntactic functions cannot be repeated in a single clause, that is, there can only be one subject, one direct object, one indirect object and so on. How can the formal structure of codons be represented in a syntax where only SPO is allowed, but not *SSP or *OOO? The codons of the genetic code are not restricted in this sense: the appearance of two similar bases (CUC = Leucine, GAA = Glutamic acid…), and even of three of them (UUU = Phenylalanine, AAA = Lysine…) is always possible. Secondly, the formal properties that characterize the functions of the clause also invalidate the target pairing: the most salient being the fact that they are signalled by agreement morphemes (in English the subject agrees with the verb), or by specific order patterns (the direct object must follow the verb), but formal patterns can hardly be said to belong to the same cognitive set as meanings (both are not 'acids').

These inconveniences would be prevented if we substituted the formal syntactic functions (i.e., subject, object. etc.) by the semantic roles that are currently distinguished by the thematic theory. From this point of view, the syntactic structure of a clause like *John shoved Mary* is not SVO, but "Agent + action + Patient", where two semantic labels are (partially) repeated since an Agent is a person that supports an action, and an action is the process supported by an Agent. Similarly, in *the child has got a cold* there are three passive notions, the Experiencer *the child*, the Object *a cold*, and the process *has got*. Moreover, these roles are obviously related to the semantic features of the clause (thus, they are 'acids'): there is an Agent because the referent of *John*, as a human being, can be the doer, etc. But unfortunately, the substitution of the formal syntactic labels, such as subject or object, by the semantic roles Agent or Patient, reopens the arbitrariness disadvantage once again: while the meaning of a clause is independent of the syntactic functions of the words the clause is made of, it is strictly dependent on the semantic roles these words fulfil.

A blind alley 89

6.3 What does the genetic code really look like and how does it work?

Perhaps the failure to establish a fruitful correspondence between the genetic code and the linguistic code is due to the naïve approach that this unusual analogy is obliged to adopt. In chapter 5 formal properties of syntax were described in such a way that it is easier than the usual description by linguists but more complex than the pedagogical (and false) image biologists are used to having of natural language as a code. I will briefly sketch the basic steps of the process by which the DNA of the nucleus of the cell is converted into protein. Once again, readers that are familiar with Genetics and Molecular Biology may pass this paragraph by.

The cell, the basic unit of life, may be prokaryotic or eukaryotic. Prokaryotic cells are simpler and more primitive than eukaryotic cells. It seems eukaryotes evolved from prokaryotes when some free-living prokaryotes were engulfed by larger cells. Prokaryotic cells (like bacteria) have very little internal organization, the genetic material is free within them. Eukaryotic cells (from protists to mammals) are structurally more complex, they consist of specialized structures (organelles) surrounded by a substance called cytosol. The genetic information is stored in the largest organelle, the nucleus (in fact, eukaryote means "with a nucleus" whereas prokaryote means "before nucleus").

Genetic information is contained in the genome, a chain of DNA molecules which are formed in the nucleus of the cell and also in the mitochondria (the energetic organelles of the cell). The genome consists of genes (30,000 genes in humans): every cell of an organism contains the entire genome, although only some genes are active, according to which tissue the cell belongs to.

Genes are made out of deoxyribonucleic acid (DNA), an extremely long polymer made from nucleotides. Four different nucleotides join to make DNA: the two purines adenine (A) and guanine (G), and the two pyrimidines cytosine (C) and thymine (T( ):

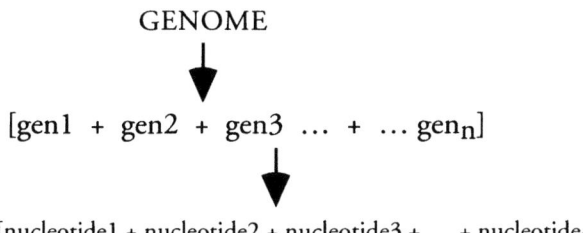

DNA molecules are very large. Humans have 46 DNA molecules in each cell, which form the so called chromosome. We inherit 23 chromosomes from each parent, which encode a complete copy of the genome. The DNA molecule is a helical polymer: this is the well known model of the double helix which was discovered by Rosalind Franklin, James Watson and Francis Crick. The model shows that the purine adenine (A) fits nicely with the pyrimidine thymine (T) forming two hydrogen bonds, and the purine guanine (G) fits with the pyrimidine cytosine (C) forming three hydrogen bonds: thus, the base pair (bp) A-T is weaker and more easily broken than the base pair G-C:

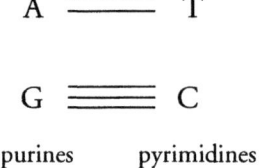

purines pyrimidines

The two chains of DNA are said to be antiparallel because they lie in the opposite orientation with respect to each other:

The information contained in DNA is transferred to its daughter molecules through replication and subsequent cell division. During replication each parent DNA strand acts as the template for the synthesis of a new daughter strand:

```
                                        daughter 1
                                        TAACTGAGGTC
                                            ▲
       <.........ATTGACTCCAG.........     ATTGACTCCAG
       ─────────────────────────────────────────────────
       .........TAACTGAGGTC.........>    TAACTGAGGTC
                                            ▼
                                        ATTGACTCCAG
                                          daughter2
```

A blind alley

Replication only happens during cell division, which is very frequent in young organisms but much less frequent as they age. However, DNA is continuously working because two other processes occur, transcription and translation.

Transcription is the transfer of DNA to RNA. Translation is the substitution of the RNA code by a sequence of amino acids during protein synthesis:

Transcription: DNA → RNA
Translation: RNA → amino acids (proteins)

Proteins are linear polymers which are made out of building blocks called amino acids. The nucleotide sequence (a sequence of bases) along the DNA chain determines the sequence of amino acids: each amino acid is specified by a group of three bases, the codon. Nevertheless, there are 20 different amino acids but only 4 different bases (A, T, C, G): the three base combinations (codons) which correspond to every amino acid constitute the genetic code.

RNA is very similar to DNA, but the former has uracil (U) where the latter has thymine: U, A, C and G are, then, the four bases of RNA. Moreover the RNA bases are tied up by ribose instead of deoxyribose which joins DNA bases together. Another important difference between RNA and DNA is that RNA consists of a unique chain of nucleotides, it is not a double helix for it does not need to be replicated. Finally, RNA loses genetic material during replication, the introns, while exons are left and newly assembled together. Only exons code for protein. Besides DNA sequences which are transcript like RNA, the human genome has a large amount of DNA which includes much repetitious DNA. The function of this DNA, whose sequence is multiplied many times, is not obvious:

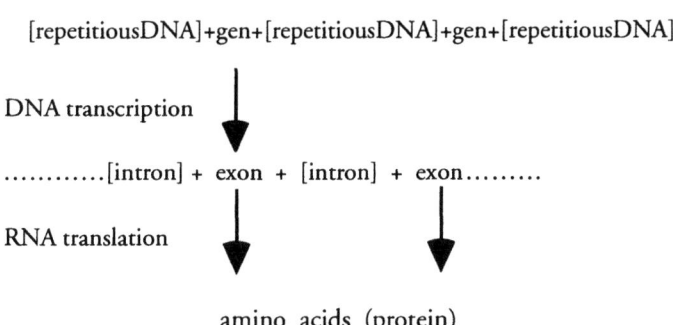

Transcription generates RNA inside the nucleus and it is a transitory process. Translation generates proteins outside the nucleus in eukaryotes and it is a permanent process.

A gene is said to be expressed when its genetic information is transferred to mRNA (messenger RNA). This is done by RNA polymerase, the enzyme that converts T, A, G, and C into the DNA template respectively in the bases A, U, C, and G in the RNA:

<............ATTGACTCCAG.........

.........TAACTGAGGTC.........>

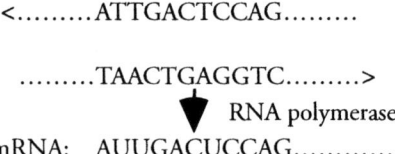

mRNA: AUUGACUCCAG............

In transcription, this enzyme must recognize the beginning of the gene to be transcribed, and must also know when it has reached the end of the gene, because genes and repetitious DNA both consist of similar chains of bases. RNA polymerase is made out of several subunits: s, b, b', and a. In fact, the role of the s (sigma) factor is to recognize a specific DNA sequence called the promoter. The a factor forms the closed promoter complex together with b or b': it separates the two DNA strands, and the RNA polymerase advances along the double helix, synthesizing an RNA chain as it goes. Finally, there are specific sequences in DNA genes, called terminators, which permit RNA polymerase to know when it has reached the end of a gene and has to stop transcribing DNA.

Protein synthesis, that is translation, is quite complex, and requires two other types of RNA, tRNA (transfer RNA) and rRNA (ribosomal RNA) to be employed. In the first step free amino acids are attached to tRNA molecules. Then, a ribosome (the translation factory consisting of rRNA), assembles on the mRNA strand and travels along it: at each codon on the RNA a tRNA binds, bringing the amino acid defined by that codon and adding it to the growing polypeptide chain:

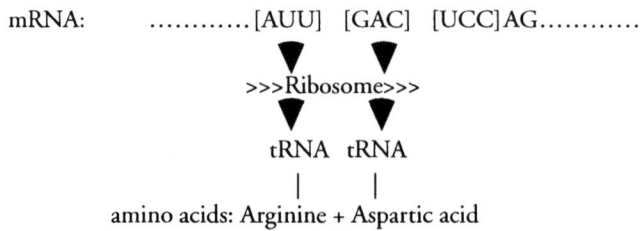

The tRNA molecules, which are a hundred bases long, have a characteristic structure called the four-armed cloverleaf. The most important feature in tRNA is the anticodon, which is a sequence of three bases in the middle of the central arm: the three bases of the anticodon bond with the three bases of a codon in the mRNA and at the same time they have a specific amino acid on the acceptor arm of the cloverleaf.

Ribosome in translation processes must also recognize specific RNA chains at initiation and at termination. But the main difference relative to transcription is that translation follows a series of coordinated processes. Organisms need to respond to changes in their environment. It would be too costly to produce all the enzymes they need permanently without using them when the substrate is absent. Consequently, complicated mechanisms developed to avoid synthesizig the enzymes of a pathway in the absence of the substrate, but guaranteeing their synthesis every time the substrate should appear. Thus, all the enzymes of a metabolic pathway are regulated together by induction and repression.

The overwhelming majority of genes in an organism code for the proteins required by the cell and are called structural genes. Structural genes are organized in clusters: regulator genes are responsible for controlling their expression:

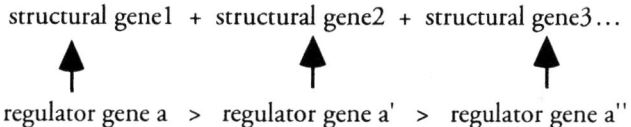

The operon, discovered by Jacob and Monod, is a unit of gene expression which includes structural genes and regulator genes. Genes controled by a negative regulator gene are transcribed unless turned off by the repressor protein that this gene codes for. Genes controled by a positive regulator gene are expressed only when an active regulator protein that this gene codes for is present.

7 On the code: the form of genetic code maps the form of linguistic code

7.1 Method: looking for a formal correspondence

The aim of the present chapter is to investigate the genetic background – if any – of the universal grammar of human language. We are looking for those features that might be considered exclusively characteristic of human language and, accordingly, those that natural selection may have stored in our brains, but not in the brains of our ancestors. These characteristics would be mainly syntactic and of a purely formal nature. A comparison of the world languages demonstrates that they share syntactic patterns, but never lexical items that could not be explained by the cultural environment (see Wierzbicka, 1996, for some exceptions that prove the rule).

The procedure we shall use is to describe a set of formal correlations between the genetic code and the linguistic one. Before any evolutionary conclusion may be drawn, we have to establish a suitable method. Since linguistic comparison is a common metaphor in any exposition of Genetics, it is important to justify the formal roots of our own pairing. We suppose that both codes, the genetic code and the linguistic, convey information. Francis Crick (1970) stated the central dogma of Genetics as a flow of information:

> The transfer of information from nucleic acid to nucleic acid, or from nucleic acid to protein, may be possible, but transfer from protein to protein, or from protein to nucleic acid, is impossible. *Information means here the precise determination of sequence, either of bases in the nucleic acid or of amino acid residues in the protein* [cursive mine].

As shown in the previous chapter, the usual approach of Figure 1 describes the sequence of nucleic acids as a series of letters, and the sequence of amino acids as a series of meanings:

```
    ... GCU  +  AGC  +  UGC ...     ...  the      train     left ...
         |        |        |                |        |        |
        Ala      Ser      Cys             "the"   "train"   "left"
```

Figure 1

Sounds are spelled in cursive, meanings in inverted commas. We have already rejected this interpretation. Nevertheless, there are some valuable aspects which we will reconsider. The most salient being the arbitrariness of the link. Notice that there is no natural link between these meanings and the sounds they are supposedly made of. Therefore, linguists call translation the substitution of the sounds belonging to the meaning of language A by the sounds of the corresponding meaning of language B, e.g., when *the train left* is translated into Spanish as *el tren partió*. Similarly, the term *translation* has been employed by geneticists to design the substitution of RNA strings (made out of codons, each consisting of a triplet of nucleotides) by amino acid strings.

But the arbitrary character of the link can be maintained under another assumption. As argued by López-García (1998), the correlation "genetic code/linguistic code" may be characterized from a productive point of view when *comparing the string of nucleic acids with the string of syntactic categories of words, and the string of amino acids with the string of their meanings*. For example, if we substitute the above string of letters that spell out the English words by the corresponding syntactic category to which each one belongs we obtain Figure 2:

```
… GCU  +  AGC  +  UGC …     … Det  +  Noun  +  Verb …
     |        |        |                |         |         |
    Ala     Ser      Cys             "the"    "train"    "left"
```

Figure 2

Now, a fundamental similarity between the genetic code and the linguistic code appears: the word *determiner* and the word *the* are both words, while G, C or U, but Ala also, are acids, etc. Actually, one of the most salient properties of linguistic code is the fact that the metacode where it is described (i.e. metalanguage) belongs to the code at the same time (López-García, 1981, 1990).

Nevertheless, although the genetic code and the linguistic code bear a close resemblance, they are not the same thing. There are two difficulties involved here:

First, each amino acid can correspond to more than one codon. For example, Ala is related to GCA, GCC, GCG, and GCU, but Trp is only related to UGG. This problem can be easily solved: *the* is only a determiner, but *book* can be both a noun (in *the book*), and a verb (in *to book*).

Second, each amino acid is related to a cluster of three nucleotides that form a codon, but linguistic words are related to single categories (metalinguistic words). This difficulty is a more serious one. It will be solved by considering the structure of linguistic phrases.

7.2 Phrase structure and Codon structure

As López-García (2002) demonstrated, we will compare the structure of the codon with the structure of the phrase. The codon consists of three components: 1st base, 2nd base, and 3^{d} base; the phrase consists of specifier, head and complement.

7.2.1 The third base and the Complement

It is a well known fact that the genetic code is a degenerated one, the type of codon being mostly unmodified by the nature of the third base. For example, no matter which nucleotide appears in the third position of Ala (whether A, G, C, or U), the resulting amino acid is completely determined by G in the first position, and by C in the second position. We are wondering whether something similar may be found in the linguistic code. Let's suppose a codon is formally compared to a phrase. As we have just said, the structure of a phrase consists of three elements, the Specifier, the Head, and the Complement. For example, in *the book of Maths*, the Specifier is *the*, the Head is *book*, and the Complement is *Maths*; in *a beautiful shirt*, the Specifier is *a*, the Head is *shirt*, and the Complement is *beautiful*. But these three components have not the same structural significance: in many phrases the Complement is not obligatory and it can be suppressed with no structural consequence at all. For example, *the book* is a phrase just like *the book of Maths*; and *a shirt* is also a phrase just like *a beautiful shirt*. Notice, however, that we could not delete the Head (e.g.: **the of Maths, *a beautiful*) nor sometimes the Specifier (e.g. *she took *book of Maths, he wears *beautiful shirt*).

In conclusion: The third base of the codon behaves formally in the same way as the Complement of the Phrase, both being structurally deletable units.

7.2.2 The second base and the Head

The nature of the second base of the codon has a structural significance. As protein structures exist largely in an aqueous environment, it is very important to know whether they will be a part of a hydrophilic or hydrophobic surface. This depends on the amino acids of their string. Amino acids can be classified on a hydrophobicity scale. Hydrophobic interactions occur between amino acids with apolar side-chains: but, interestingly enough, they are those amino acids that have U as their second base (Phe, Leu, Ile, Met, Val), and half of those that have C in that position (Pro, Ala, but neither Thr nor Ser). Hydrophilic interactions are characteristic of polar amino acids which react with water: they are those that have A in the second position (Tyr, Gln, Asn, His, Lys, Asp, Glu), and most that have G (unlike Trp, all the others: Cys, Arg, Ser, Gly).

We can thus characterize two fundamental groups of amino acids according to the nucleotide of the second position of the codon: hydrophilic amino acids that react with elements in the environment and have some relationships with them; and the hydrophobic amino acids that do not react with elements of the environment. Turning now to the linguistic side of our formal comparison, we can also characterize two fundamental types of phrase: relational phrases and constitutional phrases. This distinction can be easily understood by comparing a sentence with a web: in that structure, there are some phrases that need not be supported by any other phrase (constitutional phrases), and some phrases that cannot be conceived of without referring to other phrases of the network (relational phrases). For example, in *my sister ate the apple*, the noun phrases *my sister* and *the apple* refer to identifiable and independent entities of the external world; on the contrary, we cannot mentally represent to ourselves the verbal phrase *ate* without referring to the other two entities, someone that eats (*my sister* in this case) and to something that is eaten (*the apple*). But the syntactic nature of a phrase in the clause it forms a part of largely depends on the kind of head: if the head is a verb, or an adjective, the phrase will probably be a relational one; if the head is a noun, or a preposition (which supports a prepositional phrase), it is very probable that the phrase be a constitutive one.

In conclusion: the second base of the codon behaves as formally as the Head of the Phrase does, both either determining structures that relate to other units (polar amino acids like relational phrases), or determining structures that exist on their own (apolar amino acids like constitutional phrases).

7.2.3 The first base and the Specifier

Unlike the second base, the first base does not determine the functional behaviour of the codon. No matter which nucleotide appears in first position, the codon may be polar or apolar: UMN (where M, N are nucleotide variables) specifies apolar amino acids like Phe, Leu, Trp, and polar ones like Tyr, Cys, Ser, etc. However, the first and the third position are very different from each other in the genetic code. While the first base cannot be changed without altering the nature of the structure as a whole, the third base many times varies freely: for example, UGC is Cys, but AGC is Ser, while AGU is also Ser. This pattern significantly reminds us of linguistic specifiers: we cannot delete or change them, but they do not determine the linguistic nature of the phrase. For example, we can say *I saw the bottle* or *I saw a bottle*, but *I saw *bottle* or *I saw *some bottle* are both forbidden.

In conclusion: the first base of the codon behaves like the Specifier of the Phrase, both being significant from a structural point of view, but not being able to specify the nature of the whole.

7.2.4 Summary: a codon is formally similar to a Phrase

The fundamental similarity of both structures, the codon and the Phrase, may then be stated in Figure 3 as follows:

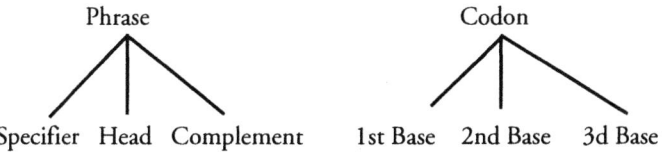

Figure 3

Nonetheless, an extended comparison would yield a rather odd picture. This is due to the fact that a phrase does not necessarily consist of three elements, because a phrase that only consists of a noun, like *Jill*, is also a noun phrase. Moreover, a phrase containing a Noun as its head is said to be a Noun Phrase, a phrase containing a Verb in head position is a Verbal Phrase, and so on. This means that in the linguistic code the entire phrasal unit behaves as the head alone does: we can say *(she likes) music*, *(she likes) that music*, or *(she likes) that music of her husband*, and the direct object of the three sentences is always a *noun phrase*. Nothing like this can be found in the genetic code, of course: GUA is a codon (Val), but neither GU nor U are codons; nor is GUA an U-codon in the same sense as "Determiner + Noun + Prepositional Phrase" is a Noun Phrase.

However, the property that supports the functional behaviour of every codon is the second base, as we said before. Whatsmore, its specifity also depends on that base. This can be easily proved when looking at Figure 4 where the codons that correspond to an amino acid, unlike Ser, always share the second nucleotide:

AMINO ACIDS	CODONS
Ala	GCA, GCC, GCG, GCU
Cys	UGC, UGU
Asp	GAC, GAU
Glu	GAA, GAG
Phe	UUC, UUU
Gly	GGA, GGC, GGG, GGU
His	CAC, CAU
Ile	AUA, AUC, AUU
Lys	AAA, AAG
Leu	UUA, UUG, CUA, CUC, CUG, CUU
Met	AUG
Asn	AAC, AAU
Pro	CCA, CCC, CCG, CCU
Gln	CAA, CAG
Arg	AGA, AGG, CGA, CGC, CGG, CGU
Ser	AGC, AGU, UCA, UCC, UCG, UCU
Thr	ACA, ACC, ACG, ACU
Val	GUA, GUC, GUG, GUU
Trp	UGG
Tyr	UAC, UAU
stop	UAG , UAA, UGA

Figure 4

It could certainly be argued that these amino acids (unlike Leu, Ser, and Arg) also share the first nucleotide of their codon, though it does not determine the behaviour of the whole. But in the linguistic code most of the specifiers are likewise head-specific ones: *the* always precedes a noun, *may* always accompanies a verb, and so on.

An additional difficulty seems to prevent the desired parallelism since a codon necessarily consists of three nucleotides, all but two of them, or even each of the three bases, may be repeated: GGA and GGG are also codons (Gly), as if we had G2A or G3 respectively. However, this picture resembles those phrases that consist of only one single constituent such as a proper noun: the formal equivalent of G3 would be then a noun phrase like *Mary* in *Mary came yesterday at noon*. This fact permits us to carry out the comparison a little bit further: every phrase could be considered as a Head with some adherences (the Specifier and the Complement), and likewise the codon would be considered as a second base plus two other added bases.

7.3 Nucleotides and categories

From a grammatical point of view there are two kinds of closely related concepts which differ slightly in meaning: the function and the category. For example, in *the white house*, the word *house* is the head, as a function, but it is a noun, as a category. This also happens to be the case in the codon: in UAC (Tyr), we have an A, as a category, that occupies the 2nd position, as a function. The head of a phrase is not necessarily a noun, it can be a verb, as in *(he) speaks slowly*: the 2nd base of a codon does not have to be an A, it can be an U as in AUG (Met).

The question now is how many categories can be distinguished in grammar and what kind of formal correspondence they keep with the nucleotides of the genetic code. There is semantic, morphological, and phonological evidence in support of the idea that sentences have a categorial constituent structure. There is even stronger syntactical evidence that supports this assumption. Developed in Chomsky (1970), the four major word-level categories Verb, Adjective, Noun, and Preposition can be analysed as complexes of two binary syntactic features, namely [±N] and [±V]. This implies a feature based analysis of syntactic categories as follows:

Noun (N)=[+N,-V] Verb (V)=[-N,+V] Adjective (A)=[+N, +V] Preposition (P)=[-N, -V]

But as argued by Radford (1988), this feature analysis allows us to define a supercategory as a set of categories which share a subset of features, as in Figure 5:

Supercategory of [+V] categories, comprising V and A
Supercategory of [-V] categories, comprising N and P
Supercategory of [+N] categories, comprising N and A
Supercategory of [-N] categories, comprising V and P

that is:

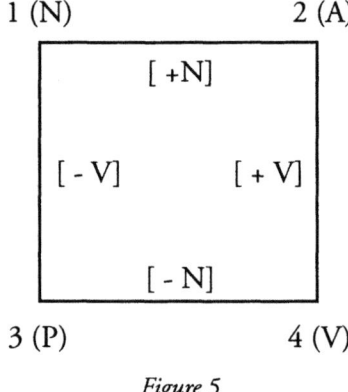

Figure 5

It seems that such a system of supercategories can be motivated on universalist grounds since natural languages behave according to these patterns. Now, one wonders whether this picture could be formally justified by the genetic code. Suppose we put the feature based foundations of this analysis in brackets. If we did so, we would have four main categories, call them 1, 2, 3, and 4. These categories would be related to each other in the four fundamental ways of Figure 6:

I, that comprises 1, 2
II, that comprises, 3, 4
III, that comprises 1,3
IV, that comprises 2,4

that is:

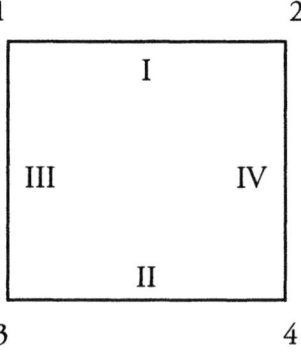

Figure 6

Turning now to the biological side, from a purely formal point of view this is exactly what happens in the genetic code. There are four nucleotides, call them 1 (Adenine), 2 (Thymine), 3 (Guanine), and 4 (Cytosine). These nucleotides are related in the following chemical ways: two hydrogen bonds link A and T in the double helix; three hydrogen bonds link G and C in the double helix; A and G are purines: T and C are pyrimidines:

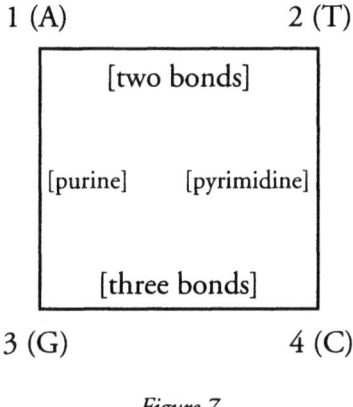

Figure 7

Call I the being-linked-through-two bonds property, call II the being-linked-through-three-bonds property, call III the being-a-purine property, and call IV the being-a-pyrimidine property. Then, we obtain Figure 6 again.

Notice that genetic sets I, II, II, and IV simply mean that the members of every set share some chemical property in common. Similarly, the linguistic sets are established according to the distributional syntactic properties shared by their members.

8 Further formal parallelisms between genetic code and linguistic code

8.1 A provisional summary of formal resemblances

I would like to summarize the points we have covered until now. Genetic Code (GC) and Linguistic Code (LC) – which, from the point of view of universal grammar is, properly speaking, a syntactic code – share the following fundamental properties:

a) GC consists of triplets of units at base level (codons made out of nucleotides) which are read as a single unit of the higher level (amino acids);

a') LC consists of triplets of units at base level (strings made out of specifier, head and complement morphemes) which are read as single units of the higher level (phrases);

b) The chemical nature of the units of both levels of the GC is essentially the same, they are acids (nucleic acids contain Phosphorus together with Hydrogen and Carbon, amino acids have Nitrogen).

b') The grammatical nature of the units of both levels of the LC is essentially the same, they are categories. The X-bar syntax is based on a two-level theory of categories, namely, word-level categories (noun, verb, etc) and phrase level categories (noun phrase, verb phrase, etc.), although further arguments should extend it to include a third type of category somewhere between these two.

c) The central unit of the string of linear nucleotides that constitute every codon of GC, namely the second nucleotide, is responsible for the biochemical behaviour of the entire triplet.

c') The central unit that supports the structure of every phrase of LC, namely, its head, is responsible for the behaviour of the entire phrase.

The concordances above between GC and LC are noteworthy, but they could be due to the general structural properties of the process that gives rise to a new level that emerges from an older one, or they may even be there merely by chance. Without doubt, the condition that one of the elements in a higher unit should determine the nature of the whole compound does not always hold: masons put bricks together to make walls without a particular brick being the head of the wall; algorithms that underlie computer programmes assemble digital characters together and none of them in particular determine the nature of the instruction they are codifying. Sometimes, however, this restriction does hold: the arches of the buildings are supported by a single central piece that takes the pressures of the other pieces. It would be more difficult to find other formal combinatorial requisites, for example, that the elements gathered together have to be only three, or that they have to posses the respective formal properties that characterize the specifier, the head and the complement of a phrase. In any case, the previous resemblances between GC and LC, as impressive they seem at a first glance, are still too scarce to allow us to draw a conclusion. Further research is required.

8.2 Satellite DNA as a formal pattern for constituent structure

Many languages are configurational. Configurationality is the propriety that relates linguistic items in terms of whole-part relationships. For example, in English, the Specifier-Head-Complement components of the NP *many tall workers* are related as in Figure 8:

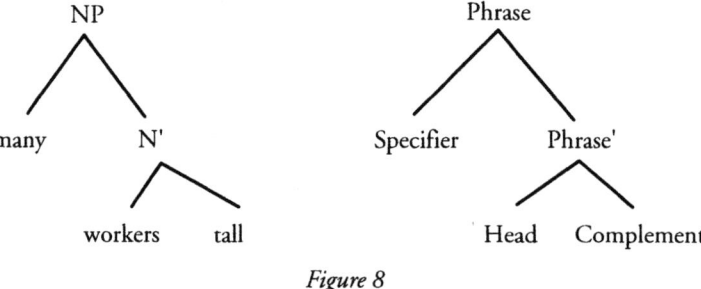

Figure 8

where *tall workers* is a 'small' NP (an N-bar, as linguists say) which is included in the major NP *many tall workers*.

There are several arguments that support such an analysis, among them the fact that *tall workers* can occur as an independent constituent in other types of sentence structure, as in *we are looking for tall workers,* or the fact that it can be coordinated with a similar sequence, as in *do you prefer tall workers or short ones?* Languages vary with respect to their grade of configurationality. Some years ago, linguists thought that the configurationality parameter opposes some languages that are configurational and have hierarchical structure to others that are nonconfigurational and have a flat structure. For the time being, we know that configurationality is rather a gradual propriety that languages always manifest.

When looking for a formal partner of the constituent structure in the genetic code, it is important to keep in mind the gradual nature of configurationality. If its genetic image belonged to the code, which is universal, then it would turn out that a propriety manifested by world languages would correlate approximately with another propriety that any organism exhibits as a whole. For this reason it seems preferable to look into the satellite DNA, which varies greatly from one organism to another.

A cell contains a large amount of DNA: there are some sequences that code for proteins, the genes, and long sequences between the genes, which are referred to as intergenic DNA. The biological function of the latter DNA, that is sometimes called 'junk' DNA, is by no means clear. Anyway, these sequences that are outside the coding regions could have a structural role. We know two fundamental classes of noncoding DNA whose structural patterns have been established: satellite DNA and transposons. Satellite DNA consists of repetitive DNA, taking the form of short sequences that are repeated, back to back, or in tandem, many times over in identical or related copies in the genome of many mammals. According to their length, we distinguish three fundamental types: satellite DNA (which in humans consists of a basic 171 base pairs sequence), minisatellite DNA (whose lengthy repetitive sequence ranges from ~15 to 100 base pairs long), and microsatellite DNA (that contain no more than two, three, or four bases). Mammalian satellite DNAs are constructed from a hierarchy of repeating units, each of them repeated in tandem on its own, as in Figure 9:

Level 1/2
... GGACCTGGAATATGGCGAG... (1–19)
... GGACGTGGAATATGGCAAG... (119–137)

Level 1/4
... GGACCTGGAATATGGCGAG... (1–19)
... GGACGTGAAAAATGACGAA... (177–191)

Level 1/8
... GGACCTGGAATATGGCGAG... (a1)
... CGACTTGAAAAATGACGAA... (a2)

Figure 9

Notice that formally these tandem repetitions of satellite DNA form a graph structure, a mathematical pattern of dependences, as shown by Figure 10:

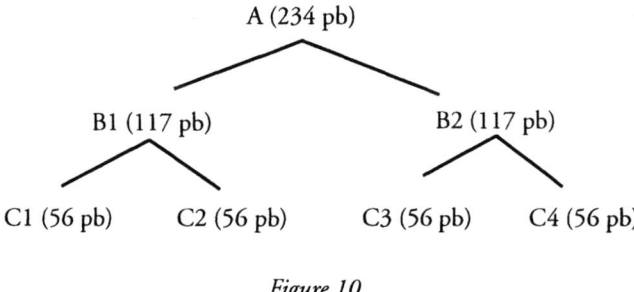

Figure 10

that resembles linguistic graphs of constituent structure. This powerful resemblance is due not only to the fact that both structures are tree-like, but also to the common internal nature of the tandem branches within each repetition. For example, on level 1/2, the DNA string between bp 1 and bp 19 is similar to the string between bp 119 and bp 137, despite the fact that the former has C in fifth position where the latter has G, and the former has G in the seventeenth position where the latter has A. Similarly, *tall + workers* may be considered two correlative tandem repetitions that belong to the same level, where *tall* consists of the cluster of features [+dimension, +person], and *workers* consists of the cluster of features [+person, +goal directed effort], which are both quite similar and somewhat different at the same time. The more we go down the hierarchical tree the less its branches are related. In the genetic tree, for example, the two halves (117 bp each) of a 234 bp sequence on level 1/2 differ at 22 positions, corresponding to 19% divergence, while within every 117 bp unit we can recognize two further subunits whose divergence has increased to 23 out of 58 positions, or 40%. In the syntactic tree the selective restrictions behave in a similar way (López-García, 2002, 9.1).

8.3 Genetic crossing-over resembles syntactic movement

A very common phenomenon of language is that of two syntactic structures consisting of (almost) the same units, and only differing in the order of their respective lexical items: *she will come tomorrow/ will she come tomorrow?; it seems to me that Mary is glad/ Mary seems to me to be glad*, etc. Linguists are used to saying that these structures are the result of movement rules: for example, starting with *she will come tomorrow*, we "move" the *will* morpheme from the second position of the sentence to the first to obtain *will she come tomorrow?* Nevertheless, it is not obvious why we should start with the declarative sentence and not with the interrogative one: if we had began with *will she come tomorrow?*, then we would have moved the *will* morpheme to the second position obtaining *she will come tomorrow*. It has been argued that vocalic contraction of *have* down to a vowel is blocked by the presence of a gap between the pronoun and *have*: for this reason, it is assumed, we cannot contract *we* and *have* through the gap that was occupied by the moved *will*, that is *will*we've finished by four o'clock?* However, this argument depends on very specific properties of the English language which are missing in some dialects. It would be much easier to speak of "correlative structures": *she will come tomorrow* and *will she come tomorrow?* correlate with each other and have merely interchanged some morphemes. This allows us to extend such an analysis to typical stylistic pairs of moved structures like *Elisabeth is my sister* and *my sister is Elisabeth*, as seen on the right in Figure 11:

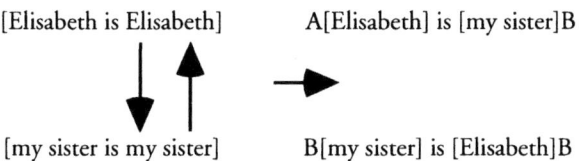

Figure 11

Surprisingly enough, the genetic code has also correlative structures that formally approach these movement phenomena of natural languages. Morgan suggested that the production of recombinant classes can be equated with the process of crossing-over that is visible during meiosis. Crossing-over consists of a chiasma, an event described as a breakage and posterior reunion: as seen in Figure 12, two of the chromatids in a bivalent structure have been broken at corresponding points, and the broken ends have been rejoined crosswise:

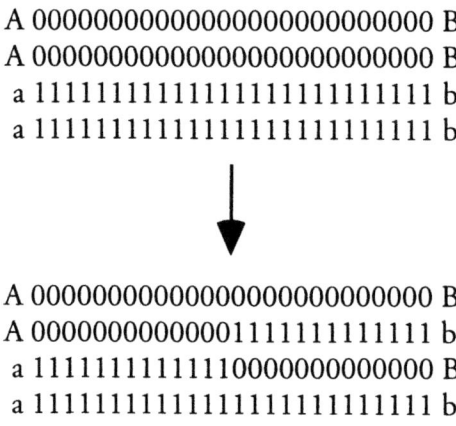

Figure 12

Each new chromatid consists of material coming from the old chromatid A-B on one side of the junction, and of material coming from the old chromatid a-b on the other side. Similarly, linguistic correlative constructions behave as if they had converted synthetic constructions of a tautological nature (*Elisabeth is Elisabeth*), which have no informative value and stored in the brain, into analytic ones (*Elisabeth is my sister*).

8.4 Crossover fixation as a formal model of recursion

Natural languages are creative, they allow us to generate an infinite set of sentence structures with a defined set of rules, in order to verbalize the infinite situations of everyday life that people confront worldwide. The formal property of language that makes creativity possible is recursion, the fact that one clause can be indefinitely embedded inside another many times: *the man that lives in the town that is in the country that was visited by the girl that is a friend of the neighbour that* […] Recursion is a possibility, not a fact: real sentences are actually not of an infinite length, but they could be, and the linguistic code must be designed in order to formally support this property.

Recursion is a formal property that also characterizes genetic crossover fixation. This is the process by which one repeating unit takes over the entire DNA satellite. Unequal recombination has been proposed to explain biological recursion. The different repeating units are closely enough related to one another to mispair for recombination. Then a series of unequal recombination events is responsible for one unit spreading around. The spread of a unit B inside a satellite consisting initially of sequence abcde, where each letter represents a repeating unit, is illustrated as in Figure 13:

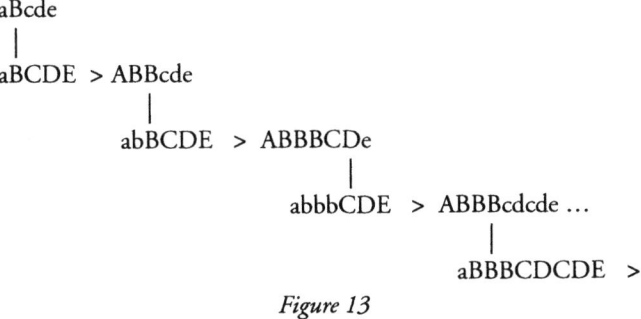

Figure 13

Once again we cannot answer the fundamental question whether this surprising correlation between the genetic code and the linguistic code has any kind of evolutionary significance, or is only a casual one. Anyway, the formal structures of both recursive patterns are very alike and therefore closer attention is needed.

8.5 Wobble as a formal schema for agreement

Till now, we have considered some formal properties of syntax that allow the speaker to structure the syntactic relationships that function within the text, and, conversely, the general cognitive pattern that allows the listener to decode the message. Let's suppose that we have taught Ameslan to a chimpanzee and it gesticulates with signs that correspond to *woman*, to *children* and *to kiss*, or, simply, that the ape points to their referents in the scene that the sentence represents. By doing so, the chimpanzee wants to communicate something to us about a woman, two children (perhaps her daughter and her son) and a relationship that is established between them. This is what Bickerton (1990) called a protolanguage, a linguistic state that our ancestors probably attained four million years ago. However, a protolanguage is not a language, and the very linguistic system does not exist until the listener is able to recognize who kisses whom, that is, whether the woman kisses the children, or it's the children who kiss the woman. Edward Sapir noticed many years ago, in a famous chapter from his well known book *Language* where he examined a similar phrase (*the farmer kills the duckling*), that many grammatical concepts may be left out in a text, but we can never dispense with either the basic material concepts (type I: 'woman', 'child', 'to kiss'), or the pure relational concepts (type IV: concepts as 'subject of' or 'object of'). These concepts, Sapir (1921, V) says, are universal and must be grammatically represented within any language.

The question now is whether the formal resources we have considered above are enough to characterize these pure relational concepts or not. At a first glance it would appear that they are. When comparing the X-bar structure of I: *the woman kisses the children* with that of II: *the children kiss the woman* (determiners are left out for the sake of simplicity):

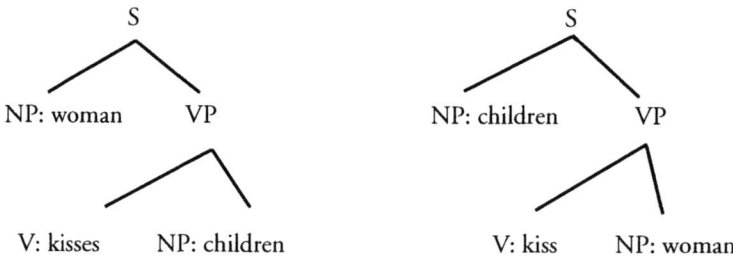

I: the woman kisses the children II: the children kiss the woman

it stands to reason, as Chomsky has repeatedly emphasized, that the 'subject of' is the NP that is immediately dominated by S (namely, *woman* in I and *children* in II), and that the 'object of' is the first (i.e. leftmost) NP immediately dominated by VP (namely, *children* in I, and *woman* in II). Moreover, Chomsky claims, relational information of this kind is entirely redundant.

However, structural trees, though justified from a cognitive point of view (and even from a genetic one, as we have seen) cannot be directly perceived by the addressee. The French linguist L. Tesnière (1959, 19–22) noticed that the left and the right positions in the tree are structural ones, and that the listener only disposes of the linear sequence to recognize the 'subject of' and the 'object of' relationships. This sequence is easy and gives them some useful indications in a chain that only consists of three items, as the strings we examined above, but it fails to reveal all the structural dependences that hold together phrases when it exceeds this dimension, as linguistic sentences usually do.

Consider the following sentence: *she was very fond of her friends and they were very fond of her too*. The speaker probably thought it up in association with a structural tree like this (once again, many formal features have been neglected):

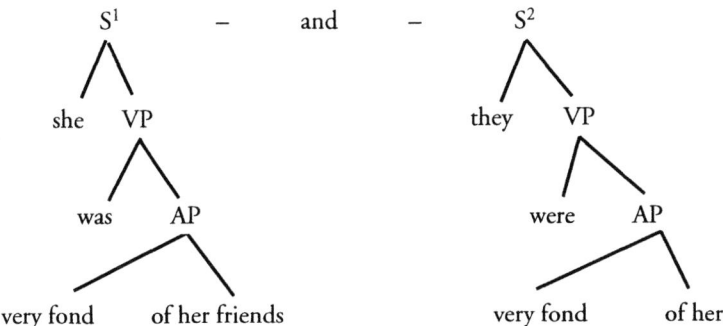

But the listener is not provided with such rich information, they only receive a linear string of morphemes: *she + was + very + fond + of + her + friends + and + they + were + very + fond + of + her + too*. How can they become aware that the group [*friends and they*] does not constitute a unit?: because the preceding string would be rather odd, something like *she was very fond of her* in spite of *she was very fond of herself*. The listener, then, analyzes the linear string and infers that in S^1 *her* is a possessive that modifies *friend* and that agrees with *she*, they are her friends. On the contrary, the listener is prevented from interpreting the morpheme *her* of S^2 as a possessive because it does not agree with the subject of this sentence, the pronoun *they*.

We may conclude that the listener is able to reconstruct the structural tree of dependencies the speaker has in mind when producing a sentence, because the former benefits from their awareness within the relationship agreement in the sentence. These relationships are manifested by means of linear morphemes, that is, within the superficial structure of the sentence (the listener takes sides), and not within the deeper structure. It is important to notice that, although languages broadly vary in the way they manifest the concordance, the agreement relationship as such always consists of a reorganization of the categorial links defined by the code. For example, whereas the four linguistic categories are related in the way V-A, N-P, N-A and V-P, which respectively share the [+V], the [-V], the [+N], and the [-N] feature, the agreement that holds between the subject and the verb of *the woman kisses the children* links N and V, that is, N-V, along the diagonal (1–4) of the square above (p. 102). Similarly, the 2–3 relationship of the other diagonal of the square allows an adjective to get linked with a preposition: this way the prepositional phrase *of her friends* is related to the subject *she* through the adjectival predicate *very fond* the former agrees with, and the prepositional phrase *of her*, that agrees with this very adjectival phrase, is referentially opposed to the subject *they*.

Despite the discussion so far, however, much of the interest in agreement is not with this grammatical property as such but with its genetic relative: what kind of formal behaviour of the genetic code resembles the selective linkage procedure represented by the agreement morphemes of the linguistic code? The proposal I would like to make here is that the agreement linkage reminds us very much of the genetic wobble. The reason is that wobble constitutes precisely a selective linkage procedure of the genetic code, and it has to this effect some relating nucleic bases that do not normally bond together.

A codon-anticodon pairing takes place when a triplet of the mRNA chain is converted in an amino acid by means of the transfer RNA. The tRNA molecules have a characteristic structure called a four-armed cloverleaf. The anticodon is part of the cloverleaf and it consists of three bases which always carry a specific amino acid. The pairing is based on the structural complementarity of bases: the first base of the codon pairs with the third base of the anticodon, and so on: codon ACG pairs with anticodon CGU in the form:

    Codon: A C G.........>
    Anticodon: <.........U G C

But whereas the pairing between codon and anticodon always follows the usual rules at the first and at the second codon positions (third and second anticodon positions respectively), exceptional wobbles may occur at the third codon position (the first anticodon one) because of the conformation of the tRNA loop. The new rules for recognition of the third base of the codon allows pairing between G and U (T in the double helix of DNA) apart from the usual pairs. As a result of this, in the first position of the anticodon U (T) is paired either with A or with G at the third position of the codon, and G is paired either with C or with U (T), while the pairings C-G and A-U act as follows:

Codon: A C G/A.........>   Codon: A C C/U.........>
Anticodon: <...U G U   Anticodon: <.........U G G

The formal consequence of these readjustments is that G and U (T) are joined along the diagonal side 2–3 of the genetic square, which gives rise to a partial redefinition of the genetic categorial pairings in the same sense as agreement does with respect to the linguistic ones. Once again, the question here remains whether or not the former pattern constitutes a model for the latter.

8.6 Empty categories follow the formal model of transposons

Sometimes linguists are forced to postulate the existence of empty categories. When a lexical category undergoes ellipsis, what happens is that it is stripped of all its lexical content and survives only as an empty node. For example, in the sentence *the boss wants to leave tomorrow*, the subject of the subordinate clause is co-referential with the subject of the main clause, and for this reason it is substituted by an empty category (*e*), i. e., *the boss wants that the boss leaves tomorrow* > *the boss wants e to leave tomorrow*. If the subject of the subordinate clause had been a non co-referential noun phrase, then the infinitive would not have appeared, and no empty category would have been postulated: *the boss wants you to leave tomorrow*. Some scholars tend to explain the appearance of an empty category as a consequence of co-referentiality. This was the case in the example above, but there are other syntactic structures where co-referentiality is compatible with formal overt categories. For example, the sentence *Mary is not sure whether she should leave tomorrow* alternates with the sentence *Mary is not sure whether to leave tomorrow*, which contains an empty category in the subordinate clause: *Mary is not sure whether e to leave tomorrow*.

The conclusion we may draw from the above considerations is that empty categories are not triggered by co-referentiality, but rather by the infinitive: every time an infinitive appears in the subordinate clause, it is incompatible with an overt subject, and the subject has to be dropped and substituted by an empty category. However, ellipsis does not only affect noun phrases. Whatever phrase appears twice in a sentence must be omitted, irrespective of the syntactic nature of its head: for instance, a verbal phrase like *my husband promised me to join the meeting, and probably he will e*, or an adjectival phrase, as in *I think her skirt was blue, but my wife is not sure it was e*.

After having left aside both suitable triggers for empty categories, co-referentialiy and the infinitive, a third possibility can be examined. Suppose that the structure of *John prefers to stay at home* is not something like *John prefers e to stay at home*, as it is usually analyzed in grammatical literature, but, on the contrary, something like **prefers John to stay at home*. In this case, since English does not allow the existence of subjectless finite clauses, the lexical subject of the subordinate clause must move to the front position of the sentence, and we will obtain *John prefers e to stay at home*, as required:

[......] prefers John to stay at home
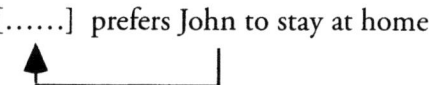

Such an analysis fits the examples considered above quite well since it does not depend on a specific grammatical trigger like the co-referentiality or the infinitive, and it is valid for any lexical category. Besides there is another reason for adopting it. Generative grammarians usually recognize two kinds of empty categories, these are symbolized as PRO and pro. PRO is the empty category examined so far. With respect to pro, it corresponds to a structure similar to the implicit subject of the Romance languages, which can be left out because of the flexive morphemes that manifest it in the verb. Spanish *quieres pan* translates into English as *you want some bread*, and the grammatical meaning "you" is represented by the *-es* morpheme of the verb *quier-es*: as a consequence of this, *quieres pan* is grammatically analyzed in the form *e quieres pan*. No direct correspondence has been found until now between PRO and pro. However, given this, it is easy to represent both in a very similar manner for pro is obtained also by putting the personal morphemes of the verb a step forward, although without deleting them in the original position at the same time:

/2nd person sing/ quier-es pan
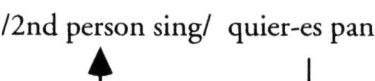

Since the personal subject morphemes are never deleted, they can either be represented optionally by means of a lexical pronoun as in *¿tú quieres pan?* (*tú* = 'you'), or left unexpressed:

tú/ø quieres pan

And once again there are some formal processes of the genetic code that behave in a similar way. The second fundamental type of noncoding DNA we mentioned above are transposons. The genome is usually regarded as something quite static that only changes through evolution, but this picture is false. Together with broad regions of the genome that only vary because of occasional mutations or recombinations taking place, there are also some permanent dynamic forces for changes called transposons. Transposons are sequences of DNA that are able to move from one site to another. Usually each transposon carries genes that code for the enzyme activities required for its own transposition. In this sense, transposons are autotransposable elements. Most of them are also considered as 'selfish' DNA for they are concerned only with their own propagation and confer neither advantage nor disadvantage to the phenotype. This gives us a closer view of pro, an empty category that is optionally manifested as a lexical item.

It is very remarkable the way transposons work. The simplest transposons, called IS (insertion sequences), consist of a series of base pairs in which its ends are identified by inverted terminal repeats, while the site of the host DNA is identified by a short direct repeat which appears once again, after the IS insertion, in its original form. For instance, a landmark TTGCA of the host (M representing any base: A, T, G, or C):

Host DNA: ...MMMMMMMM TTGCA [target site] MMMMMMMM...

attracts an insertion sequence situated further on in the DNA string (N representing also any base):

IS: AATGAGCTG (inverted repeaat) NNNNNNNN (transposon) CAGCTCATT (inverted repeat)

which is flanked by two corresponding inverted repeats, the left one AATGAGCTG, and the right one CAGCTCATT; the latter is formed by those bases that fit the corresponding bases of the former when disposed in the reverse order (remember that A is joined with T, and G with C):

AATGAGCTG
TTACTCGAC

The resulting structure looks as follows:

...MMMMMTTGCA [AATGAGCTG NNNNNNNN CAGCTCATT] TTGCA MMMMM...

The subject of the infinitive behaves like the transposon: a slot in the subject position of the main clause attracts it, although the force to be moved comes from the lexical item itself when it accompanies an infinitive which is unable to agree with it. The transposon is a formal unit defined by the two inverted repeats that surround it: similarly, the lexical unit that moves from the subordinate clause to the main one must not be a subject and has only a formal character (for example, it is an object in *they are anxious for me to leave*).

Two kinds of transposition are habitually envisaged in the genetic literature: replicative transposition and conservative transposition. The last one follows the above scheme, and reminds us of PRO, as depicted in:

conservative

XXXX [...] XXXX YY XXXX   as [...] prefers John to stay at home

The first one, on the other hand, consists in duplicating the insertion sequence, so that the transposing entity is a copy of the original, which remains in its place. Replicative transposition elucidates the formal mechanism of pro:

replicative

XXXX [YY] XXXX YY XXXX   as tú [2nd sg] quier-es [2nd sg] pan

8.7 Subjaceny is framed like cis-dominance

As we will see in the next chapter, some DNA sites are able to control other genes. However a distinction arises depending on whether the action of these DNA sites is restricted only to adjacent genes or extends, by means of diffusible products, to any other genes irrespective of their location:

The phenomenon of *cis-dominance* means a DNA sequence that does not specify any product (usually a protein) that can diffuse through the cell and exercise its effect outside the same DNA molecule;

The phenomenon of *trans-dominance* means a DNA sequence that specifies a diffusible product, hence acting on all relevant sites in the cell, whether they are present on the same or different molecules of DNA.

Now, when considering the linguistic side, we are tempted to relate cis-dominance with locality conditions. A DNA site is cis-acting when it functions by being recognized inside a specific region whose limits cannot be surpassed. Similarly, there are some linguistic movements that cannot go beyond a syntactic boundary. For example, the contrast between *how do you think John got it?* and **how do you wonder whether John got it?* demonstrates that it is much more difficult to extract a wh-phrase out of a clause that has a wh-phrase in its complement position than it is from a clause which has not. In these cases linguists tend to speak of syntactic islands which are specified by a single condition called subjacency: movement cannot cross more than one bounding node (IP and DP) in a single step. The crossover to the genetic side is easy: DNA molecules act in cis-dominance like bounding nodes do in linguistic subjacency.

9 Linguistic texts and genomic strings share some formal devices

9.1 On genetic levels and linguistic levels

Every time biologists employ the term *genetic code*, they are used to understanding three different notions:

i) The set of formal correspondences that underlies the substitution of each one of the 64 codons by one of the 20 amino acids. It is strictly speaking: we examined this aspect in chapter 4.
ii) The rules that govern the substitution process in practice. We have examined this facet in chapter 5 by considering the failures that occur in the translation, and also those DNA strings that are not translated at all.
iii) The association of amino acids in proteins, which is regulated by some formal charachteristics of the DNA string. This question will be considered in the present chapter.

Perhaps it would have been better to use the term *code* to designate the first aspect only. Nevertheless, the polysemy of the word *code*, as registered by the Webster's dictionary is responsible for the extension above, since it defines *code* (s. v.) as:

I) A comprehensive and systematically arranged body of law;
II) A system of rules of conduct or procedure;
III) A system of signals used in communication.

A body of law (I) makes sense only when these laws are employed in practice (II), and its general purpose being that of transmitting some instructions to some receivers, it follows that the code above has to materialize in the form of some concrete messages (III). The genetic code i generates some DNA sequences ii, and these sequences are read as polypeptidic chains iii. A striking parallelism with the linguistic code manifests once again. Linguistic code consists of an inventory of categories i' that are combined together by means of a system of rules ii', and the resulting product of such an operation are texts iii'. As systemic linguistics stated some fifty years ago, linguistic code is based upon the notion of rank: for each language there are a small number of syntagmatic slots that are singled out for the most effective description of the paradigmatic relations in the language.

However, some difficulties arise when we consider the successive levels that are attained in the construction of a text. Firstly, some syntactic categories are tied up in a phrase; secondly, phrases are joined together to form sentences; thirdly, some sentences that convey a common meaning constitute a text. For example the text:

All children, except one strange little boy, grow up one day. Wendy knew she would have to grow up when she was just two years old (Peter Pan)

consists of the sentence:

All children, except one strange little boy, grow up one day

and of the sentence:

Wendy knew she would have to grow up when she was just two years old

Besides the first sentence can be analyzed on its own as the result of adding the phrase *except one strange little boy*, and, afterwards, the phrase *grow up one day*, to the phrase *all children*, whereas the second sentence can be seen as the sum of *Wendy knew*, of *she would have to grow up*, of *when she was*, and of *just two years old*, although this linear splitting up seems insufficient because it does not consider the X-bar structure.

Similarly, biologists distinguish three successive levels of DNA composition, the codon level, the cistron level, and the operon level:

```
codon:    ...AAN GUC CAU CAC UUA AUG GCN...
am.ac:    ...Lys Val His His Leu Met Ala...
cistron:         polypeptide 1              +    polypeptide 2
operon:          protein 1
```

As a conclusion, we can tentatively propose the following parallelism:

LINGUISTIC CODE	GENETIC CODE
phrase	codon
sentence	cistron
text	operon

We indicated in a previous chapter that a codon is a group of three successive base pairs that are read as an amino acid. A cistron is the functional notion that corresponds to a gene, that is, a string of codons responsible for coding the amino acids that constitute a polypeptide, the minimal unit that is able to trigger some biologic reaction. Finally, an operon is a set of genes/cistrons that are regulated by the same gene/cistron. The problem in maintaining this correlation between LC and GC lies in the extension of their respective units. On the one side, codons always consist of three base pairs; phrases may, ranging from about seven to one, but formally they are also three, the specifier, the head, and the complement. On the other side, although some texts may include only a single word (as in *stop!*, etc.), usually linguistic texts are made out of several hundreds, and even thousands, of lexical categories (words), and this is also the case with the number of operons bases. Nevertheless, when we compare the extension of sentences with the extension of genes/cistrons, they do not overlap: a linguistic sentence usually consists of a dozen words (it can even consist of only one word), while a gene/cistron has on average one thousand bases or more.

Probably this quantitative divergence between the gene/cistron and the sentence is rather apparent. In fact, any sentence contains much more than the meaning it literally expresses. The difference between language and genome is that linguistic sentences are heavily polysemic, they have to be emitted as utterances. And every utterance communicates a variety of propositions, some explicitly, others implicitly. As Sperber and Wilson (1986) pointed out, the meaning of an utterance includes not only what the speaker intended to say, but also what he/she intended to imply, and the speaker's intended attitude to what was said or implied. In its turn each one of these enlarged meanings is decomposable to several elementary propositions, according to its logical form. The result of also considering these atomic propositions and those non literal meanings that underlie every sentence is that the number of phrases are drastically increased. This is why artificial intelligence needs to implement the formal representation of sentences with extensive systems of explanation as the ATT-Meta system (Barnden, 1998) does.

9.2 Sentences as transcriptional units

We have just claimed that the difference between sentences and cistrons lies in the fact that sentences are much shorter. Where do the unexpressed meanings go? This question cannot be answered without distinguishing the role of the speaker from that of the listener: every time the speaker conceives a message, they builds up the sentence as a set of cognitive structures most of which remain unexpressed until the listener rediscovers them through understanding. The set of meanings that constitute a sentence are then cut off and only a small subset is expressed: this is the utterance.

The conversion of sentences into utterances formally resembles the genetic transcription of DNA into RNA. In prokaryotes (organisms without a nucleus) every three bases in the DNA chain are translated as an amino acid. There are eukaryotes in most organisms – humans included –, the DNA is lodged in the nucleus (although there is also mitochondrial DNA, within the cytoplasm). But translation, the process by which codons are read as amino acids, takes place in the cytoplasm. This means that DNA must travel from nucleus to cytoplasm. The process is as follows: firstly, the totality of a gene's DNA is transcribed into preRNA (precursor RNA); secondly, the genes are split into alternating RNA segments, those that code for amino acids, the exons, and those intervening sequences that do not get an polypeptidic interpretation, the introns; thirdly, the exons are joined together in the same order that they appear in the DNA to form mRNA (messenger RNA); forthly, the mature mRNA is then transported to the ribosomes of the cytoplasm (the factory of translation) for protein synthesis.

A gene/cistron is, then, a transcriptional unit of the form "… intron + $exon_i$ + intron + $exon_j$ + intron…" that, after splicing, is converted in "…$exon_i$ + $exon_j$…". Similarly, a sentence is a linguistic unit within the speaker's mind that consists of several propositions and that, after loosing some of them during the emission of the message, it is converted into an utterance. The propositions that are lost remain implicit and can be later restored in the listener's mind; the propositions that are expressed by the utterance are the explicit ones: "… implicit proposition + explicit $proposition_i$ + implicit proposition + explicit $proposition_j$ + implicit proposition…" > "… explicit $proposition_i$ + explicit $proposition_j$ …". The parallelism between genetic transcription and linguistic emission would read as follows:

genetic transcription	linguistic emission
GENE	SENTENCE
...intron+exon+intron+exon+intron...	...implicit+explicit+implicit+explicit+implicit...
↓	↓
...exon+exon...	...explicit+explicit...
mRNA	utterance

This parallelism is not based on a mere superficial analogy. For example, the way mutations alter exons is different to the way they affect introns, and a corresponding inequality is found in propositions. Exons, because they direct the construction of proteins, are much more highly conserved in evolution, any change would alter the entire coding pairings and produce a disastrous effect. However, introns are not indifferent to mutations: they do not impede the expression of proteins, but only hinder it to some extent. This is also the case with the linguistic partners of exons and introns: any change in an utterance alters the meaning of the explicit proposition that the speaker intended, as when the speaker says *there is a cat* instead of *there is a rat*; however, the listener may not understand *there is a rat* in any figurative sense (like in *you dirty rat!*), but in a literal one, and ask themselves what the animal is: this would be a change in an implicit proposition.

9.2.1 The promoter and the talk openings

Parallelisms between genetic transcription and linguistic uttering of sentences go a step further when we consider the structure of both processes. Genetic transcription is catalyzed and scrutinized by the enzyme RNA polymerase, but it does not begin at any point in the DNA string. On the contrary, there is a specific sequence of DNA, called the *promoter*, that is needed for the initiation reaction. However, the enzyme initially fails to find the promoter. The initiation stage is protracted by the occurrence of abortive initiation, when the enzyme makes shorts transcripts and aborts them. This occurs repeatedly, until the RNA polymerase succeeds in moving to the next phase. The enzyme chooses the promoter because the so called sigma factor is attracted by some consensus sequence that precedes it. For instance, in *Escherichia coli*, a bacterium of human intestine, the typical promoter has the following form:

TATAAT + 16/19 irrelevant bp + TTGACA + 5/9 irrelevant bp+ initiation site

The generalized consensus sequence is actually TATAAT ... TTGACA. Its discontinuous structure prevents the sigma factor being confused with any other stretch of DNA of the genome. If, after TATAAT is found, the enzyme does not find TTGACA (the so called Pribnow box) approximately 17 pair bases afterwards, the initiation is aborted, and the process begins again.

The linguistic utterance of a participant does not begin at any point of a given speech act. As a matter of fact, talk displays an overall organization with clear beginnings and carefully organized closings. There are generally some marks that allow the participant to take a turn. Topic selection plays a fundamental role in this practice: when someone participates in a discourse recognizes that a new topic has been introduced, they adapt their utterances accordingly.

Conversational structure was first analyzed by Schegloff (1972) based on telephonic interaction, most of its formal characteristics are also valid for any other types. Openings always consist of a three-turn structure, T_1: *summons* – T_2: *answer* – T_3: *reason for summons*: the third turn introduces the *first topic slot*. For example, many telephone conversations have as their first three turns the following, or something very similar:

(summons)	RINGS
(answer)	*Hello*
(reason for summons: greetings)	*Hello Bob. This is John. How is everything?*

Such a three-turn structure is not exclusive to telephone conversation. For instance, a typical face to face opening could be:

(summons)	*Mary?*
(answer)	*Yeah?*
(reason for summons)	*Could you pass the wine, please?*

Notice that genetic promotion of transcription and conversational openings share formal structure. There is always a discontinuous consensus sequence, TATAAT ... TTGACA in the genetic transcription, *summons – answer* in the speech act. However, abortive initiation frequently occurs: in the linguistic code this is the case when the phone rings and no one picks it up or when people call someone but do not receive any answer; in the genetic code it is the situation we described above when the sigma factor fails to recognize the Pribnow box. The two first turns do not immediately follow each other: the two parts of the consensus sequence are separated by 16 / 19 bases; similarly, there can be some pieces of discourse by any participant except the addressee between the two members of the adjacent pair *summons – answer*, as in:

John:	*Mary?*
John (to Peter):	*I can't find her*
Mary:	*Here I am*

After the opening section consisting of these two turns, what is called the *first topic slot* usually appears with an announcement by the speaker of the reason for the call. In the linguistic code this is the reason for summons turn, in the genetic code it is the initiation site. Both of them may be preceded by some irrelevant sequences (for example, *How are you?* [*Could you give me a piece of advice?*]).

9.2.2 Control at termination and the talk closings

It is difficult to define the termination point of the transcription process. Once the enzyme has been started by RNA, polymerase moves along the DNA-RNA hybrid until it meets a terminator site. At this point the enzyme stops synthesizing nucleotides adding them to the RNA chain, which is dissociated from the DNA template. In spite of initiation, where the promoter is announced by a typical consensus sequence, no formal structure characterizes termination. As a consequence of this, *in vivo* a terminated string and a string that has been cut look alike. Experiments made *in vitro* have demonstrated that, according to whether RNA polymerase requires any additional protein to terminate, two devices may be distinguished in *E. coli*:

Rho-independent terminators: no additional protein is needed;
Rho-dependent terminators: the addition of a protein, the rho factor, is needed.

Rho-independent terminators stem from the combination of two formal structures: hairpin generated by pairing between corresponding bases of the DNA which form a loop; and a string of ~6 U residues at the very end. The hairpin causes the polymerase slow or pause. Then, the run of U residues, whose association to the corresponding A string requires the least energy of any RNA-DNA hybrid to break, provides the signal that allows RNA polymerase to dissociate from the template, although sometimes this U tail may be missing.

Rho-dependent terminators work when this protein is present at about 10 % of the concentration of the RNA polymerase. The process progresses step by step as follows: first, RNA polymerase transcribes DNA; second, Rho attaches to recognition site on RNA; third: Rho moves along RNA until it catches up with the polymerase at terminator; forth, Rho unwinds DNA-RNA hybrid and all components are released.

Consider now the issue of termination at the linguistic side. When does a particular sentence finish? In front of its beginning, which, as we have seen, is always recognizable because there are some formal devices that help to identify it, no specific marks announce that a particular sentence will be the last one in a given speech act. After having heard a sentence like *we decided to fix the old house up ourselves*, we can make two claims about its continuation: perhaps a detailed explanation about the things to be repaired in that house will follow, maybe the speech act will stop precisely at this sentence. However, there are some points where the speech flow cannot stop in any case: for instance, we could not interrupt it after *to fix*, that is **we decided to fix*, for this verb is a transitive one and the information it carries has to be completed. Unfortunately such an argument is only valid within the limits of the clause the sentence consists of, but it does not allow us to predict any possibility outside it. Nevertheless, many times people are aware that a speech act has exhausted its possibilities and that the time to close it has arrived. The clue to it is generally the fact that any new information the listener might obtain from the speaker about the subject has already been manifested. The information is organized along the discourse in the form theme-rheme: the theme comprises the given information, the rheme is anything new. Every time the flow of discourse arrives at a new piece of information, that is, at a rheme, it slows down. But finally it arrives at a particular rheme that completes the inquiries that the theme had made: this is the final stop. The structure of discourse ending closely resembles that of the rho-independent terminator:

theme+rheme$_1$ (pause) > theme+rheme$_2$ (pause) > ... > theme+rheme$_n$ (FINAL PAUSE)

like

bps+ hairpin$_1$ (pause) > bps + hairpin$_2$ (pause) > ... > bps + hairpin$_n$ + UUUUUU (FINAL PAUSE)

and the role of the ~6 U tail is played by some optional discourse endings like the tags that introduce a new turn (*He said you like working in the factory, don't you?*).

There is not a unique formal device for finishing the discourse. Sometimes the closure of a sentence is triggered by the focus and not by the last rheme of a series. The focus is one of several alternatives. After having manifested the others, the discourse ends when the missing one manifests itself. For example, in *Some of the Gothic churches turn me on, but some of the Renaissance palaces turn me off*, the appearance of the verb *to turn on* announces the possibility of a partner like *to turn off*. This is a neutral focus. A contrastive focus is a functional alternative that is absolutely opposed to any other one and is high pitched, for example *this car does not belong to my sister, but TO MY BROTHER*. Both patterns ressemble the Rho-dependent termination: like the Rho protein a linguistic mark appears first, and the sentence ends when its lexical counterpart (*off* in relation to *on*, *brother* in relation to *sister*) finally manifests.

9.3 The utterance markers and the translation markers

Transcription has been completed when polymerase reaches the above termination sequence at the end of a gene: then it detaches from the DNA, and the translation process begins. The completed mRNA moves from the nucleus to the ribosomes, where the polypeptide chain is assembled. Protein synthesis is achieved by transfer RNA molecules. These tRNA molecules have a characteristic structure called a four-armed cloverleaf. Its most important feature is the anticodon, which is a sequence of three bases that is capable of bonding with the three bases of a codon in the mRNA. Besides the anticodon, which is complementary to its respective codon, each tRNA has an amino acid. This way, amino acids are assembled together in a chain, and a polypeptide is formed.

Notice that translation is the process by which a cell reacts to the requirements of the environment. While DNA or RNA remain inside the nucleus, the genetic code is still a latent possibility. Only after "it knows what is needed", can the cell react by transforming some codons of nucleic acid into a product of a different kind: the amino acid. Language works in a similar way. Sentences are formal patterns that remain asleep in memory until the requirements of the verbal context compel them to convert in utterances. An utterance is the linguistic equivalent of a chain of amino acids. As we said above, the difference between codons and amino acids is not that of 'sound vs. meanings', as claimed by the classical metaphor, but rather that of 'language vs. parole', 'emic vs. etic', or 'deep vs. surface structure'. Unfortunately, although we know quite well how the genetic translation works, the process that converts a mental sentence into the utterance of a speech act which takes place in the speaker's brain, despite some psycholinguistic endeavours, is mostly unknown.

However, some formal similarities between both processes hold. The mRNA does not move directly after transcription from the nucleus to the cytoplasm. Before this, it receives some protective marks against enzyme attack. These marks are the so called *cap* (a methylated compound), at the beginning of the mRNA, and a poly A tail at the end. Besides, all gene chains are preceded by an initiation codon, AUG, and followed by one of three termination codons (UAA, UAG, UGA), all of which are meaningless codons. In conclusion, the structure that enters the ribosome in the cytoplasm is of the form:

cap + [consensus sequence with Gs and Cs] + AUG + codons + UAA / UAG / UGA + poly A tail

The question now is whether sentences also need to be protected by some kind of wrapping when they convert themselves in utterances. And the answer, surprisingly enough, is once again: yes. For a sentence to be an utterance it has to be provided with an intonational pattern. Sentences exist as propositions with a grammatical structure in the mind, but they only adjust to a specific environment when they turn into utterances. The intonational contour does not have a meaning on its own, but it is required for a sentence being converted into an utterance. The sentence pattern is characterized especially by the toneme of the first and last syllable of the utterance. The formal parallelism with the genetic code is obvious: every gene has to be preceded by the initiation codon AUG, and followed by a termination codon, which although meaningless, are obligatory in order to decode the gene. Sometimes, however, these stop codons receive an interpretation, and this is also the case in languages like Spanish where a high toneme at the end of the utterance has an interrogative meaning, and a low toneme has a declarative one. Besides these intonational obligatory contours that envelope any utterance, sometimes there are initial or final pauses, which remember respectively the cap and the poly A tail. The whole picture looks like this:

cap + [consensus sequence with Gs and Cs] + AUG + codons + UAA / UAG / UGA + poly A tail
(pause) + first toneme + utterance + last toneme + (pause)

9.4 The regulation of texts and the operon

The issue of the interconnections between the sentences in a piece of discourse has recently been discussed in great detail in the linguistic literature. There are very deep meaning connections that relate to the encyclopaedic knowledge of the world that the speaker and the listener share and without which no understanding of the text is possible. The genome works in a very different manner, it has only a single polypeptidic interpretation and it resembles much more the string of instructions followed by the machine language of a computer. This aspect of the problem does not provide a suitable basis for comparison in any case. But on the surface level things are much closer in both domains, in the linguistic text and in the genetic text, that is, in the operon.

It is obvious that sentences which constitute a text cannot be put in any order whatsoever. For example, as pointed out by Hoey (1983, 4–6), the following sentences do not conform a coherent and understandable text:

(1) In England, however, the tungsten-typed spikes would tear the thin tarmac surfaces of our roads to pieces as soon as the protective layer of snow or ice melted.
(2) Road maintenance crews try to reduce the danger of skidding by scattering sand upon the road surface.
(3) We therefore have to settle for the method described above as the lesser of two evils.
(4) Their spikes grip the icy surfaces and enable the motorist to corner safely where non-spiked tyres would be disastrous.
(5) Its main drawback is that if there are fresh snowfalls the whole process has to be repeated, and if the snowfalls continue, it becomes increasingly ineffective in providing some kind of grip for tyres.
(6) These tyres prevent most skidding and are effective in the extreme weather conditions as long as the roads are regularly cleared loose snow.
(7) Such a measure is generally adequate for our very brief snowfalls.
(8) Whenever there is snow in England, some of the country roads may have black ice.
(9) In Norway, where there may be snow and ice for nearly seven months of the year, the law requires that all cars be fitted with special spiked tyres.
(10) Motorists coming suddenly upon stretches of black ice may find themselves skidding off the road.

The order in which these sentences originally appeared was 8-10-2-7-5-9-6-4-1-3, that is:

(8) Whenever there is snow in England, some of the country roads may have black ice.
(10) Motorists coming suddenly upon stretches of black ice may find themselves skid-

ding off the road. (2) Road maintenance crews try to reduce the danger of skidding by scattering sand upon the road surface. (7) Such a measure is generally adequate for our very brief snowfalls. (5) Its main drawback is that if there are fresh snowfalls the whole process has to be repeated, and if the snowfalls continue, it becomes increasingly ineffective in providing some kind of grip for tyres. (9) In Norway, where there may be snow and ice for nearly seven months of the year, the law requires that all cars be fitted with special spiked tyres. (6) These tyres prevent most skidding and are effective in the extreme weather conditions as long as the roads are regularly cleared loose snow. (4) Their spikes grip the icy surfaces and enable the motorist to corner safely where non-spiked tyres would be disastrous. (1) In England, however, the tungsten-typed spikes would tear the thin tarmac surfaces of our roads to pieces as soon as the protective layer of snow or ice melted. (3) We therefore have to settle for the method described above as the lesser of two evils.

Significantly enough, when Nunan (1993, § 1.1.1) asked a group of undergraduate students to restore it, the results demonstrated close agreement as to what was an acceptable ordering. This demonstrates that there are some formal mechanisms that allow us to link sentences in a text, and that they belong to our implicit knowledge of the linguistic system together with the syntactical and the morphological resources any native speaker knows by heart. However, such mechanisms are not reduced to anaphoric of several kinds as Hoey argues:

black ice (8), black ice [...] skidding (10), skidding [...] scattering sand on the road surface (2), such a measure (7), its (5), tyres (9), these tyres (6), their [...] spikes (4), spikes (1), method described above (3).

It is easy to imagine another string of anaphoric expressions, that relate to each other, but that do not constitute a coherent message, for example, 9-5-3-1-10-4-8-7-6-2:

(9) In Norway, where there may be snow and ice for nearly seven months of the year, the law requires that all cars be fitted with special spiked tyres(h). (5) Its main drawback(i) is that if there are fresh snowfalls the whole process has to be repeated, and if the snowfalls(j) continue, it becomes increasingly ineffective in providing some kind of grip for tyres(h). (3) We therefore have to settle for the method described above as the lesser of two evils(i). (1) In England, however, the tungsten-typed spikes would tear the thin tarmac surfaces of our roads(k) to pieces as soon as the protective layer of snow(j) or ice melted. (10) Motorists(l) coming suddenly upon stretches of black ice may find themselves skidding off the road(k). (4) Their spikes grip the icy(m) surfaces and enable the motorist(l) to corner safely where non-spiked tyres would be disastrous. (8) Whenever there is snow(n) in England, some of the country roads(o) may have black ice(m). (7) Such a measure is generally adequate for our very brief snowfalls(n). (6) These tyres prevent most skidding and are effective in the extreme weather conditions(p) as long as

the roads(o) are regularly cleared loose snow. (2) Road maintenance crews try to reduce the danger(p) of skidding by scattering sand upon the road surface.

This text is nonsense, although it satisfies the formal anaphoric requirements stated above for every sentence repeating some lexical item of the sentence(s) that precede it, as the subscripts (h, i, j...) manifest. We agree with Nune when he argues that in addition to linguistic knowledge (that is, knowledge of how sentences are formed internally, and combined with each other externally), we also need to be aware of the subject matter of the text, of what he calls non-linguistic knowledge. Nevertheless, this type of knowledge may be dispensed with in most narrative texts where the writer tells a story that the reader did not know before.

Although non-linguistic knowledge, that is, encyclopaedic information, is obviously very important and should not be disregarded by the analysis, the textual coherence of a string of sentences lies mainly in themselves. Let us examine the coherent text again. First we will underline the verbs:

(8) Whenever there is snow in England, some of the country roads may have black ice. (10) Motorists coming suddenly upon stretches of black ice may find themselves skidding off the road. (2) Road maintenance crews try to reduce the danger of skidding by scattering sand upon the road surface. (7) Such a measure is generally adequate for our very brief snowfalls. (5) Its main drawback is that if there are fresh snowfalls the whole process has to be repeated, and if the snowfalls continue, it becomes increasingly ineffective in providing some kind of grip for tyres. (9) In Norway, where there may be snow and ice for nearly seven months of the year, the law requires that all cars be fitted with special spiked tyres. (6) These tyres prevent most skidding and are effective in the extreme weather conditions as long as the roads are regularly cleared loose snow. (4) Their spikes grip the icy surfaces and enable the motorist to corner safely where non-spiked tyres would be disastrous. (1) In England, however, the tungsten-typed spikes would tear the thin tarmac surfaces of our roads to pieces as soon as the protective layer of snow or ice melted. (3) We therefore have to settle for the method described above as the lesser of two evils.

Notice that verbs are generally not repeated: each sentence contains a verb on average, and, most important, a verb that neither appeared before, nor will appear later. Now we will put in italics many nouns of the model text A:

(8) Whenever there is *snow* in England, some of the country *roads* may have *black ice*. (10) *Motorists* coming suddenly upon stretches of *black ice* may find themselves *skidding off* the road. (2) *Road* maintenance crews try to reduce the *danger* of *skidding* by scattering sand upon the *road* surface. (7) Such a measure is generally adequate for our very brief *snowfalls*. (5) Its main drawback is that if there are fresh *snowfalls* the whole

process has to be repeated, and if the *snowfalls* continue, it becomes increasingly ineffective in providing some kind of *grip* for *tyres*. (9) In Norway, where there may be *snow* and *ice* for nearly seven months of the year, the law requires that all cars be fitted with special *spiked tyres*. (6) These *tyres* prevent most *skidding* and are effective in the extreme weather conditions as long as the *roads* are regularly cleared loose *snow*. (4) Their *spikes grip* the *icy* surfaces and enable the *motorist* to corner safely where *non-spiked tyres* would be *disastrous*. (1) In England, however, the tungsten-typed *spikes* would tear the thin tarmac surfaces of our *roads* to pieces as soon as the protective layer of *snow* or *ice* melted. (3) We therefore have to settle for the method described above as the lesser of two *evils*.

So far, scholars agree (Halliday and Hasan, 1976) that the coherence of discourse is due to some formal devices that help to tie the sentences in a text together, such as reference, substitution, ellipsis, conjunction and lexical cohesion. Many of them may be found in text A. However, we might conceive of a text B where some of the lexical items of A also appear, where they refer to the same entities of the world, where formal connexions reveal that their sentences belong to the same set too, and, notwithstanding, B would exhibit a very different meaning coherence:

> Motorists are disastrous for black ice stretches of the roads. When passing over, they break them down, so the lovely landscape made out of snow and ice suddenly disappears under their spiked tyres. A solution could be that the law requires that they skid off the road. Their vehicles had to be fitted with non-spiked tyres in order to prevent them of gripping the roads.

What happened? We put together a rather clumsy text B, that means just the opposite meaning of A, by employing almost exactly the same words. The reason lies in the fact that lexical items were maintained, but thematic relations were violated. For example, motorists, that in A are the patients that suffer inconvenience by icy roads, are converted in B to agents that destroy the ice. Similarly, roads, that in A were an object that had to be conserved by maintenance brigades, are now viewed in B as a location where the people, unable to grip, skid off.

We may conclude that, besides a set of conjunctions and a sample of lexical items, which are language specific, the textual coherence seems to be determined, on universal grammatical grounds, by two formal conditions:

1) The maintenance of a web of lasting thematic roles through the text;
2) The succession of constantly new predicative nodes, which link those thematic roles together, throughout the text.

Both conditions must be satisfied at the same time. A text whose thematic roles change, or that are attributed to other referents, is a new text; for instance, if Romeo were no longer in love with Juliet, that is, if he had turned himself into a passive element, or if another participant of the plot, say Tybalt, were her partner, then the drama would never have been a well known piece of William Shakespeare. But the dramatic action must advance at the same time in order to maintain the isotopy of the text; this is the reason why the death of Romeo and Juliet also means the end of the drama itself, no further predicate being conceivable after their disappearance.

9.5 Positive and negative control

In the case of genetic code, if we were to take this type of textual considerations, we would need to draw attention to the so called *induction* and *repression* processes. In a cell few genes are individually transcribed, most are part of larger units, which look like a cluster of genes, that is, like a genetic text: the operon. This allows all the genes of the cluster to be coordinated by the interaction of a regulator protein, which usually is located close to the promoter of the sequence. The *regulator protein* controls the transcription by binding to particular sites on DNA and it is coded by a *regulator gene*. Such a mechanism is fundamentally explicable on economic grounds. An organism has to respond swiftly to changes in its environment. However, the response needs to be adequate, and in proportion to the importance of the environmental change. Survival may depend on the capacity to switch from metabolizing one product to another, but, at the same time, the organism has to save energy expending only what is strictly necessary.

Biologists recognize two formal devices for the organic regulation of the expression of structural gene clusters by gene regulators:

1') *Positive control*: genes under positive control are expressed only when an active regulator protein is present;
2') *Negative control*: genes under negative control are expressed unless they are switched off by a repressor protein.

We are tempted to compare positive control to the maintenance of thematic roles, and to compare negative control to the succession of predicates in a text.

9.5.1 Positive control

The isotopy of a given text is guaranteed by a specific web of thematic roles. As long as they are unchanged, that is, as long as each one of the noun phrases of the text maintains its thematic nature, this text is said to be the same text. For example, if there is a couple who work hard to get food for their children, we can imagine a story of the troubles they have to survive until the children grow up and leave home. The plot remains structurally unmodified: NP^1 (the couple) is the Agent, and NP^2 (the children) is the Patient, and NP^3 (the food) is the Theme, etc. No changes are expected until a new web of thematic roles appears. Usually, the referents of the new thematic role cluster are already hidden in the previous text, or derive implicitly from it: for instance, there was a young boy in the neighbours house that sometimes played with the children of the couple; when this neighbour falls in love with the daughther, a new situation arises and the novel deviates from its original narrative line to tell us the story of the young couple. When this happens, thematic roles refer to a different situation in the world, and we speak of a new text.

This reminds us of the mechanism, called induction, which controls the route of lactose in a bacterium called *Escherichia coli*. When glucose is available, *E. coli* uses it in preference to other sugars as an energy source, and this preference prevents the expression of the genes of the lactose operon. But when the regular provision of glucose decreases, the lactose of the environment activates the lactose operon allowing the bacteria to choose an alternative metabolic pathway. From then, *E. coli* obtains the glucose it needs synthesizing the lactose. No induction of the operon is possible without a sufficient amount of lactose in the environment.

At a first look it may seem that the comparison of 1 (the maintenance of a thematic cluster in a text until it is broken by a new cluster) with 1' (the metabolization of glucose until it exhausts and lactose induces a new pathway) is an extravagance, or even nonsensical. In any case we could define it as a loose analogy, of only pedagogical interest, that absolutely lacks scientific power. However it is interesting to compare not only the general picture of both processes, but also the formal details of each one. The operon of the lactose works as follows:

REGULATOR GEN + PROMOTER + OPERATOR + STRUCTURAL GENES → Lactose
↓
repressor

While consuming glucose the regulator gene of E. coli produces a repressor which associates with the operator and blocks the production of lactose. However, when the level of glucose decreases, the lactose of the environment induces the inactivity of the repressor. The repressor gets rid of the operator, and then the promoter initiates the production of lactose, which consists of a sequence of connected polypeptides: b-galactosidase, coded by lacZ; b-galactosidase permease, coded by lacY; and b-galactoside transacetylase, coded by lacA.

The coherence of a given isotopic text also consists of a series of anaphoric pronouns that are successively related to each other: *John$_i$ asked Mary$_j$ a glass$_k$ of beer and she$_j$ gave it$_k$ to him$_i$ before he$_i$ could get it$_k$ by himself$_i$*. But this sequence of linked elements (i: *John > him > he > himself*, etc.) depends on the maintenance of a specific thematic cluster. When a new cluster appears, for example, *Peter had bought the beer in Susan's grocery*, the precedent series of phoric pronouns shed their referents, and a new series begins with a specific pronominal grid (*he$_i$ likes to buy her$_j$ beverages because there$_k$ are the best prices of the town*). Now, we are able to relate this mechanism to the one described above. In every case, there are two environmental products that compete with each other: glucose and lactose in the metabolism of *E. coli*, the referential web *John – glass of beer – Mary*, and the referential web *Peter – beer – Susan's grocery* in the linguistic text. While the first one prevails, its anaphoric pronoun grid continues to be tied up with the text and blocks the appearance of the second one. However, when the first referential web decreases in activity and importance, then its anaphoric grid gets rid of the text, and the second referential web imposes its own anaphoric grid which articulates a new semantic pathway, that is, a new textual orientation.

9.5.2 Negative control

Negative control in genetic operons can be exemplified by the route of Tryptophan in *E. coli*. The bacteria synthesize this amino acid through the action of the enzyme tryptophan synthetase. But if Tryptophan is provided in the medium or if the level of Tryptophan produced by *E. coli* surpasses a specific quantity, the production of the enzyme is immediately halted. This effect, which enables the bacteria to avoid devoting resources to unnecessary synthetic activities, is called repression, and it works as follows:

REGULATOR GEN + PROMOTER + OPERATOR + STRUCTURAL GENES → Tryptophan
↓ ↓
repressor (inactive) .... <<<<<<< co-repressor
↓
(active)

This kind of control is the most common and it supposes auto-regulatory control. The regulator gen produces an inactive repressor, and the RNA polymerase is able to join with the promoter and initiate the transcription of structural genes. Tryptophan is then the end product of the reactions catalyzed by a series of biosynthetic enzymes. When the level of this product is too high, any further production of the repressor is prevented because Tryptophan works as a co-repressor that joins the repressor actively blocking the operator. When the level of repressor drops, the protein fails to inhibit its own synthesis, and the level is restored.

The above control procedure is very similar to the formal device that prevents the repetition of some predicates within an isotopic text. For example, consider the text next: *The weather suddenly got very cold. The temperature dropped to five grades below zero and snow fell every day. John had arranged for heating to be installed at home but it had not yet been put in. It was very cold. [John could no more bear his room. He got up and went out].* Different predicates follow, in the same sense as enzymatic products appear one after the other in the pathway of the Tryptophan: *to get cold, to drop, to fall, to arrange, to instal, to put in, to be cold* [...] However, the repetition of the predicate *to be/get cold* closes the isotopy of the text. From here, the textual direction must be reoriented and a new piece of the story is born.

Certainly, transitions are almost infinitely more subtle and they need not strictly rely on explicit lexical repetition, but on contextual one. A text may change its orientation simply because the information it contains already satisfies the informational necessities of the addressee, or because some details that could have been added are supplied by the external environment, and so on. This situation is parallel to the blocking of the Tryptophan route, not because it has produced too much amino acid, but because the amino acid is provided by the scientist in the laboratory. For instance, pay attention to the two paragraphs of this text of *The Waves* of Virginia Woolf:

> Birds swooped and circled high up in the air. Some raced in the furrows of the wind and turned and sliced through them as if they were one body cut into a thousand shreds. Birds fell like a net descending on the tree-tops. Here one bird taking its way alone

made wing for the marsh and sat solitary on a white stake, opening its wings and shutting them.

Some petals had fallen in the garden. They lay shell-shaped on the earth. The dead leaf no longer stood upon its edge, but had been blown, now running, now pausing, against some stalk. Through all the flowers the same wave of light passed in a sudden flaunt and flash as if a fin cut the green glass of a lake. Now and again some level and masterly blast blew the multitudinous leaves up and down and then, as the wind flagged, each blade regained its identity. The flowers, burning their bright discs in the sun, flung aside the sunlight as the wind tossed them, and then some heads too heavy to rise again drooped slightly.

As can be seen, the first paragraph is concerned with birds, the second with flowers. When the first topic is exhausted, it is the second one that takes the turn.

The formal device that underlies a grammatical kind of ellipsis called 'gapping' is very similar to repression. In the text *John eats meat on Saturdays and Susan [eats] fish on Sundays* it is possible to omit the second occurrence of the verb *to eat*, that is, *John eats meat on Saturdays and Mary fish on Sundays*. However, we can not have **John put Fido in the doghouse, and Sam put Spot in the yard*. Culicover and Wilkins (1984, 29–30) suggest that this is due to a restriction on gapping that reads as follows: gapping of a verb may not leave more than one complement of V within the containing V-bar. However, since the view we have adopted here, there is a complementary explanation too: *on Sundays* belongs to the environment (it is an adjunct) and imposes a change of textual perspective which does not allow the verb *to eat* to appear again; on the contrary, *in the yard* is not an adjunct, but a complement of the verb *to put*.

10 The organizer

10.1 The inheritance of formal codes

The problems we have dealt with in the preceding chapters gain a new and particular significance when they are examined under a different perspective put forward by the above conclusions. I have pointed out certain alleged similarities between syntax and genetic code. I am aware that this attempt is subject to objection that the comparison was built on rather superficial resemblances, and that no causal chain between the two codes has been suggested at all. Further efforts may improve the picture, definitely, but it is unlikely that the main argument be substantially altered. The endorsement of my approach has to be considered, then, in its current design, although it is by no means a definitive state of art.

With this in mind, remember that the fundamental parallelisms that support the claim that syntactic form of natural languages is related to genetic code are as follows:

1) The three bases of each codon are supposed to resemble respectively the specifier, the nucleus, and the complement of a phrase;
2) The second base element is supposed to behave in a fashion similar to that of a lexical category, the subcategorization restrictions the later imposes on the complement being similar to the restrictions the former suffers when constituting chemical links with the elements of its environment. In a similar way, the structural properties of the first and of the third base are supposed to resemble the structural properties of the specifier and the complement;
3) The phenomenon of genetic crossing-over is supposed to resemble syntactic movement;
4) The formal pattern of satellite DNA is supposed to resemble X-bar structure;
5) The phenomenon of cross-over fixation is supposed to resemble recursion;
6) The phenomenon of wobble is supposed to resemble agreement;
7) The phenomenon of transposition is supposed to resemble empty categories;

8) The phenomenon of cis-dominance is supposed to resemble subjacency;
9) The sentence level is supposed to resemble the cistron level;
10) The explicit meanings of the sentence are supposed to resemble the exons, the implicit ones, the introns;
11) The promoter is supposed to resemble the talk openings;
12) The rho-independent terminator is supposed to resemble the rheme;
13) The rho-dependent terminator is supposed to resemble the focus;
14) The initiation codon is supposed to resemble the first toneme;
15) The termination codons are supposed to resemble the last tonemes;
16) The text level is supposed to resemble the operon level;
17) The positive control of the operon is supposed to resemble the thematic web that helps to maintain the isotopy of the text;
18) The negative control of the operon is supposed to resemble the ellipsis, and particularly the gapping phenomenon.

In my opinion this table looks very impressive. Nevertheless, I am aware that critics of this intended parallelism may think the resemblances are weak, or too cursorily described to evaluate, or both. Unfortunately, it seems difficult to meet on a neutral point. Although critics may not completely exclude the possibility of some causal relationship between the genetic code and syntax, they would maintain that any such relationship needs to be based on much more solid evidence. How much and what type of evidence: this is a question I am unable to answer.

As a matter of fact, this is the epistemological weakness of any analogical argument whatsoever. When does a resemblance become an homomorphism and deserve scientific plausibility? Current approaches to the linguistic evolution issue are used to drawing evidence from many domains, including problem-solving abilities in apes, the songs of birds, the structure of sign languages, computer simulations of language emergence, brain impairments with linguistic effects, or the structure of current language. And the main focus here is that these endeavours are analogical too. Why had we credit all of them, but reject the point of view I have adopted in this study?

I suggest the reason could be that in those analogical approaches the non-verbal pole of the comparison was examined until no further structural patterns were found. This makes the evolution to language a mere problem of improvement: If we discover n structural patterns in animal communication, say, in the songs of birds, it stands to reason that mankind, a species that evolved from animals after having passed through many other steps, has to possess n×m structural patterns in the human language. In a similar way, a very simple computer program constitutes a suitable base for explaining how neural webs evolved in human brains until they were complicated enough to support language. Even when the starting point is language itself, argumentation proceeds from less to more complex codes: this is the case with pathological linguistic systems, or with the alleged extension of the patterns that exist on one level to the whole linguistic system. An attempt has recently been made to find a parallelism between syllable structure and basic sentence structure (Carstairs-McCarthy, 2000).

Notice that all these approaches are not at fault from either a lack of quantity, nor from a lack of quality. It is by no means a problem that they can gather only a minimum subset of the formal structures of language because we may be confident that evolution has turned them into something bigger. In fact, body structures that are related to mind also evolved following the same pattern: the number of neurons of the brain increased when mouse step converted into dog step, and again when dog step converted into chimpanzee step, and once again when chimpanzee step converted into human step. At the same time, we are not suspicious of the poor resemblance the prelinguistic structures share with the linguistic ones for the former having evolved into the latter. Without accepting that the human eye, a very sophisticated organ, has evolved from the visual organs of preceding mammals, and before that from the eyes of frogs, fishes, or even flies [...], no theory of evolution is possible, and creationism reappears as the only alternative explanation.

However, when we compare the genetic code with the linguistic code, we allow for neither quantitative nor qualitative acquiescence. The number of formal structures shared has to be high enough, and their similarity has to be strong enough: on the other hand, analogy lacks scientific power. This is certainly a reasonable side to take because genetic code is not foregoing from where humans come from, but a formal structure that coexists with linguistic structure in humans. And even: genetic code is shared by almost all living beings, while only humans possess language. No wonder the reluctance, presumably, critics will possibly show with the proposal I am currently considering!

10.2 The limits of emergence

There is no doubt that language was an emergent level in evolution. Functional approaches to language consider it as a product of culture, and culture itself as an emergent property that only humans (and, perhaps, apes to some extent) exhibit: language would be, thus, a meme. From a generative point of view language is also considered to be an emergent property, although it is rather conceived of as the result of the interactions of the neuronal network of the brain. The main argument supporting such an assumption is the same in both cases: complexity. Human language cannot be described by simply starting with the communicative skills of animals, as sophisticated and intriguing as they may well be. The naïve claim by Darwin, who presumed that language was born when natural selection slowly improved animal cries or gestures, is definitively wrong: in fact, as observed by Chomsky (2002, IV), the exhaustive comparison of animal communication systems by Hauser (1996) demonstrates that language belongs to an entirely new domain of knowledge.

Unfortunately the complexity argument is weakened by itself. A formal object which seems to be related to a simpler one is said to have emerged from it as a complex object not simply because the complex object has some units that the simple object does not possess, or because new relationships have arisen, but because the emerging level has a set of specific properties that characterize complexity. For example, body organs made out of cells, like the liver or the heart, are complex objects, but a wall made out of bricks, and even an entire building, are not.

Complexity is opposed to reductionism. The reductionist point of view in science takes us from a complex phenomenon to the more elementary properties of its components, but it is not able to take the opposite direction in order to explain how the properties of the whole come from its parts. The sciences of complexity are faced with situations where wholes cannot be described in terms of the properties or the interactions of their parts. On the contrary, complexity must be described on its own, as an intermediate situation between order and disorder. Complexity is related to chaos, unlike their respective origins: both share the unpredictable evolution of the system, but chaotic systems are sensitive to initial conditions, whereas emergent complex systems are dependent upon the inability of the observer to predict their behaviour.

The sciences of complexity are fundamentally different from classical science for they are alien to predictability and the control of nature. Complex scientists can no longer live outside the phenomena they are interested in: the order of ecosystems, the properties of organisms, the dynamic activities of communities, these are emergent systems where we humans participate actively in. Classical science managed to describe ordered patterns, those that exhibit a low-entropy set, and where a small part is formally equivalent to the whole: a piece of iron which is made out of molecules of Fe, or a wall which is made out of bricks. It opposes disordered patterns, those that exhibit high-entropy rates, and those where a small part is statistically equivalent to the whole. Complex sciences, however, are not equivalent to either low- or high-entropy: the most interesting fact about them is that relevant patterns are discovered on many different scales. This surprising property is called fractality. A fractal object is one that exhibits the same formal appearance at any observed scale: river networks, cities, planetary surfaces, etc., are fractal objects. Some formal patterns defined by mathematicians like the set of Mandelbrot are also fractal objects.

The most important conclusion we may draw from this is that the language system as a whole is not a fractal object and, presumably, that it has not emerged from other communicative systems through evolution as a result of complexity. There are, of course, some emerging (thus, complex) patterns of language like those of the X-bar theory. The typical structure that projects, in its classical version, a noun into a noun phrase, and, on a higher scale, a noun phrase into a clause, and, on an even higher scale, a clause into a sentence, etc., reminds us of a fractal structure. No matter on what scale we approach an X-bar structure, it will consist of an X which is projected into an X' which is projected into an X". This is presumably the reason why Chomsky characterized merge as an emergent property of the minimal program. But a main part of the linguistic system is not organized this way. The attempt by Berwick (1998) to derive the remaining syntactic principles as emerging properties looks rather unsuccessful: for the same reason the numbers and the chemical elements, which are said to share the merge law with language, would also have developed these very principles.

The reason for excluding emergence as a source of many properties of the linguistic system is modularity. A linguistic system consists of several modules which function relatively independent of each other and which follow a specific set of patterns and of structuring properties. For example, the syntactic module combines the projection principle (~merge) with a structural tree that organizes the maximal projections a text is made of into a coherent pattern. On the contrary, the phonological module, although it also benefits from merge, does not organize the resulting blocks in a tree-like pattern but rather in a linear one. For example, we can analyze the syllable /'mai/ as a projection of the vowel /a/, and the noun phrase *my best friend* as a projection of the noun *friend* in essentially the same manner: in both cases the head increases without loosing its categorial specificity. But only the syntactic sequence bears covert relationships that are not dependent on linearity. In the sentence *my best friend put the lights out*, the unit *my best friend* is related to a verbal phrase *put the lights out* whose verb *put* is related to the preposition *out* jumping over the unit *the lights*, etc. In the corresponding phonetic chain, however, every sound influences its neighbouring sound, and every syllable influences its neighbouring syllable: thus, [put] influences [ðe'laits] and [ðe'laits] influences the syllable [aut], and so on. A similar divergence can be found when we compare the morphological word formation module with the syntactical one: both share merge (a word consists of a root and of an affix that are grouped together), but only syntax makes use of empty categories: there would be no sense in saying that since *responding* consists of *respond + ing*, and *corresponding* consists of *co + respond + ing*, there should be an item **respondence*, manifested as *respond*, where the affix *-ence* is an empty morpheme because of the existence of *correspondence*.

The above considerations do not exclude the possibility of some linguistic module being organized as a fractal object to be born on an emergent level. As in the case on a textual level. How is a text built up? Unless it is some prototypical texts, such as laws or prayers, people do not form them by following a previous foreseeable pattern. A speaker who wants to form a sequence at the phonetic, morphologic, or syntactic level has a set of items and some formative rules at their disposal: they select some elements, and later they merge them in an upper unit. But a text, say a piece of conversation or an entire book, does not arise as the result of such a process: actually, although speakers know how to start, they completely ignore what directions it will take and what kind of components it will include. Text formation always happens on the brink of chaos, and the final product incorporates such heterogeneous components as previous texts, contextual information, emotional feelings, and so on. Moreover, its structure is typically that of a fractal object: a novel or a dramatic play consists of a plot made out of a sample of stories which, in turn, are composed of several episodes, etc.: the most important fact is, however, that the main plot, the stories, and the episodes all share a common narrative actantial structure.

For instance, consider the abstract that The *Oxford Companion to English Literature* (s. v.) has made of the play *Volpone* by Ben Jonson:

The organizer 147

> Volpone, a rich Venetian without children, feigns that he is dying, in order to draw gifts from his would-be heirs. Mosca, his parasite and confederate, persuades each of these in turn that he is to be the heir, and thus extracts costly presents from them; one of them, Corvino, even sacrifices his wife to Volpone in hope of the inheritance. Finally Volpone overreaches himself. To enjoy the discomfiture of the vultures who are awaiting his death, he makes over his property by will to Mosca and pretends to be dead. Mosca takes advantage of the position to blackmail Volpone; and Voltore, a lawyer, who has aided Volpone in the infamous conspiracy against Corvino's wife, finding himself defrauded of his expected reward, reveals the whole matter to the senate; whereupon Volpone, Mosca, and Corvino receive the punishment they merit Ö The names of the principal characters, Volpone (the fox), Mosca (the fly), Voltore (the vulture), Corbaccio (the crow), Corvino (the raven), are significant of the parts they play.

Notice that this play is hierarchically organized in three levels: 1) the habitual relationships some prototypes, which are represented by animals, maintain (a fable); 2) the main story, the way in which Volpone tricks his would-be heirs, and the way he is deceived by Mosca; 3) several secondary stories that tell us how Mosca coaxes every would-be heir. This is a fractal structure, a pattern where the same features are reproduced at all scales of observation over and over again. A text, as a fractal structure, is then a self-similar object: its properties manifest itself under any observation, and it is hence not reducible to smaller, more fundamental units. On the contrary, other domains of linguistic system are not fractal: syntactic units that follow syntactic laws are made out of words, which obey morphological laws; words that behave according to morphological rules are made out of phonemes which obey phonological laws. When going beyond the sentence level we cross a fundamental line between the domain of regularity and of modular systems to the realm of complexity and fractal systems.

10.3 The concept of pre-program

The syntax of natural language not only consists of emergent properties. Together with the complex fractal principles we examined in chapter VIII, there are a lot of features that accommodate the nature of the external world, and also some other principles that cannot be explained, neither as the result of epigenesis, nor as an interpretation of the external world: they are pre-programmed principles.

EPIGENETIC PRINCIPLES	INTERPRETABLE FEATURES	PRE-PROGRAMED PRINCIPLES
They are born as emergent properties of a complex system, starting from the perceptual level	*They reflect culture and are not inherited: languages differ from each other on the basis of these features according to the world they live in*	They are inherited, but they have not arisen as fractal objects: they rather look like a formal pre-program

What is a pre-program? The term was coined by R. Thom (1988) in order to distinguish two kinds of pregnances. We have previously examined the normal pattern of pregnance emission: a pregnant form influences a salient form and produces a figurative effect on it (pregnance > salience). But there is also a very remarkable situation where the affected element is another pregnance, that is, a pregnance emitted by a salience affects a pregnance (salience > pregnance). How is the inversion of the expected result possible?: because saliences may become structured patterns, and act as such.

A salience generally is either an obstacle that opposes to the advance of the pregnance, or a point which changes its direction. This is the case with the current of a river every time it meets an obstacle. If the obstacle is big enough say, for example, a dam, it simply stops, however, with something small, like a stone in mid-stream, the current flows around as it continues. But when the water meets a tap or a mill its salient form imposes a specific formal pattern to the resulting pregnance. The formal pattern acquired by the stream is, then, a structure that has been induced by a device: the pre-program.

According to the general definition by Thom, a pre-program is a salient form that is immersed in a stream and whose movement can produce some morphologies in the stream. Most tools are mobile pre-programs: the ax, the knife, the propeller, the pump are all pre-programs responsible for several catastrophes. For example, the blade of a knife performs the catastrophic dual cusp dividing an object in two. However, pre-programs are not necessarily related with mechanical appliances. There are also, Thom claims, biological pre-programs: thus, the dynamics of cell that characterizes mitosis is a mobile pre-program which is moved by the energetic stream of the environment; the morphogenesis of animals also follows a general plan of organization, that is, a pre-program. In these cases the pre-program manifests a successive process of bifurcation: a cell converts itself into two cells, and each of them is divided in two more cells, and so on; similarly, the undifferentiated blastoderm differentiates endoderm, mesoderm and ectoderm, etc. Notice that these dynamical processes are not of the fractal type, as observed by Thom, unlike plants which follow mathematical graph models and behave in a fractal manner, like the Fibonacci series. Animal morphogenesis does not proceed as a self-organizing (i.e., emerging) process, but it develops according to a plan of organization.

10.4 The organizer

This is where we now come across the concept of organizer. Typical vertebrate embryos consist of three cell sheets: the outer layer (ectoderm: is the origin of epidermis and nervous system); the middle layer (mesoderm: produces some organs, like heart, connective tissues, like bones or muscles, and blood cells); and the inner layer (endoderm: where the digestive tube comes from). These layers are already separated in the germinal egg even before gastrulation: the ectoderm in the animal pole domain; the mesoderm in the equatorial domain; the endoderm in the vegetal pole domain. Traditional embryology did not go any further, but when in 1924 the German biologist Hans Spemann and his Ph.D. student Hilde Mangold made a surprising discovery: biological forms resulting from morphogenesis are sometimes induced by the activity of some tissues, the so called 'organizer'.

It was a well known fact that at the beginning of gastrulation some areas of an embryo are not yet functionally determined: it is possible to exchange some parts of the ectoderm that during further development would have become neural plate and some other parts that would have become epidermis without disturbing normal development. Spemann and Mangold proved that such a transplant is feasible not only between embryos of the same age and of the same species but also between different ages or species: for instance, presumptive brain tissue of *Triton taeniatus* transplanted into the epidermal region of *Triton cristatus* can become epidermis. But a piece from the upper lip of the blastopore behaves quite differently: when it is transplanted into the presumptive epidermic region, it does not accommodate to the environment, as expected, but it develops according to its origin, that is, it develops as a secondary primordium with neural tube, notochord, and somites. Thus, the concept of an "organization centre" emerged: a region of the embryo which preceded other parts in determination and is capable of determining new areas in certain domains.

Since then, researchers have spent a lot of time at tempting to characterize chemicals that induce development. During the thirties and forties they proposed a set of neural-inducing molecules from the ectoderm so unlike each other that there appeared to be no chemical specificity: they found some acids (oleic, nucleic), some proteins, and even some non-inducing regions of amphibian gastrula that had been killed by ethanol treatment. It was Waddington (1962) who reinterpreted the entire paradigm describing induction in terms of molecular biology: chemicals that determine specific formation patterns of the embryo are induced by gene expression, the entire problem being oriented, thus, to the discovery of such genetic factors. The impressive amount of molecular analyses which runs through his book showed that primary embryonic induction, as conceived by experimental embryologists, was neither primary, nor induced: the organizer tissue itself was the product of a prior induction, and the neural fate of cells was the default, uninduced, fate of ectodermal tissues.

Modern studies have established that embryonic induction begins at the organizer of the organizer, the so-called Nieuwkoop centre, which is a region of the presumptive dorsal endoderm that induces the organizer proper in the mesoderm above it. Therefore, the organizer is responsible for pattern formation providing information about time, place, scale and orientation for the development of several groups of cells that surround it. The process has proven to be a very complex one, and new discoveries are continuously added to our knowledge of it. In relation to the gastrulation of *Xenopus laevis*, for example, the list of inducers (genes that induce gastrulation), and of their related induced regions of the presumptive gastrula has been summarized by Bouwmeester (2001) as follows:

REGION	TRANSCRIPTION FACTORS	SECRETED FACTORS
Dorsoanterior endoderm	Xhex, mix genes, Xblimp-1	cerberus, dkk-1
Prechordal plate	goosecoid, Xotx-2, Xlim-1, Xant-1	dkk-1, frzb-1, chordin, noggin
Notochord	Xnot, pintallavis, Xlim-1, Xbra, Xerg-1	chordin, noggin, follistatin, admp
Floorplate	Xfd-12', pintallavis	
Organizer epithelium		Xnr-3

A great number of genes are expressed within the organizer. This gives rise to genetic cascades and networks of interactions between genes: some of them inhibit the expression of forms and some others have an activating function. The traditional concept starting from Spemann-Mangold's organizer experiment was that embryonic tissue plays a role in the development of mesoderm and neural plate. Unfortunately, its existence could only be inferred from the results of transplant experiments. For the time being the situation looks much more complicated, but great advances have been made in our understanding of the facts. For the time being we know that the three cell layers of the embryo enter the differentiation pathway due to the action of many molecules synthesized by the organizer upon the responding tissues that surround it. In general terms, BMPs (Bone Morphogenetic Proteins) are secreted by the ventral side of the embryo and are antagonized by organizer secreted factors which bind them to the extra cellular space of ectoderm, mesoderm and endoderm. The first homeobox gene to be isolated was goosecoid (Cho et al., 1991); since then, many other genes have been described. Moreover, although Spemann only recognized two parts, now we are aware that the organizer has at least three: the head- organizer, the trunk-tail organizer, and the deep yolky endoderm. Finally, these inducers and their corresponding mechanisms have been discovered not only in amphibians, but also in Drosophila, Zebrafish, Mice and Human. Nevertheless, three fundamental conclusions remain:

Firstly, patterning is not entirely a self-organizing process, there are also some regions of the embryo (the organizer) that participate.
Secondly, organization by the organizer is preceded by the activation of a predecessor (the Nieuwkoop centre).
Thirdly, the organizer has some parts which differ in their inducers, their morphogenesis, and their self-differentiations.

Researchers have paid close attention to the significance of the organizer and respect to the timing of the egg's development. Without a predetermined pathway where specific genes are turned on and off at particular times success cannot be attained. Genes work like a cascade, a chain where every time one gene is turned on/off at one stage the expression of other genes is controlled. This produces the picture of a series of genes. But from a molecular point of view the problem is to convert a continuous gradient into the discrete differences that determine each cell type. Three types of genes are involved in this process: maternal genes, segmentation genes (which are responsible for the number and polarity of segments), and homeotic genes (which control the identity of every segment). Many of the homeotic and of the segmentation genes display motifs that are simple repeating sequences, the most common of which is the homeobox. The homeobox is a region up to 180 bases, which has remained constant in size and almost basic in sequence through evolution. As we said in § 7.2, homeobox genes are genes that contain homeobox sequences. They tend to lie in clusters. The parallel of hox clusters among mouse, fly and other animals raises the extraordinary possibility that these clusters of genes share a common evolution.

10.5 Homeobox once again

To date, it is not apparent what relationships exist, if any, between the organizer activities that manifest in morphogenesis and the molecular activities that occur inside the cell. The processes that take place within the cell have obvious repercussions on the network the cell belongs to. But, unless we are tied up with an extreme reductive approach, the former does not explain the later anyway. This assumption underlines the weakness of all strict empirist approaches: for instance, we know that thoughts are supported by brain-wired networks, and we know that these networks are supported by a set of synapses, and we know that an individual synapse is a relationship between two neurons, and we finally know that the behaviour of each single neural cell can be described in terms of DNA, RNA, proteins, etc. But even so, Logic and Philosophy cannot be reduced to Neurology, neither can this be reduced to Molecular Biology. The unique reductive couplings modern scientists have succeeded to develop since Galilean revolution were, firstly, that of unifying Physics and Chemistry in the nineteenth century, and, secondly, that of relating Biology to both of them in the twentieth century. This means that our declared purpose of extending biomolecular considerations to mental processes – and language is one of them – still looks unfeasible.

But astonishingly enough, a pattern has been discovered that directly links the intracellular phenomena with some phenomena outside the cell; a pattern then, that organizes both of them. Transcription, the decoding process that converts DNA in RNA, is performed in eukaryotic cells by three classes of genes which are defined by their respective types of promoter. However, these promoters, although they are transcribed by three kinds of RNA polymerase (I, II, and III), are not recognized by the polymerase without the help of the so called transcription factors.

Transcription factors are proteins that are needed for the initiation of transcription, but that are not themselves part of RNA polymerase. Usually they recognize cis-acting sites of the promoter, or recognize another factor. The sequences (the TATA box, GC box, CAAT box, etc., among them) of the target promoter are bound by the transcription factor which, in turn, may bind polymerase or other transcription factors. Although binding affects the entire transcription factor, it is performed by quite short 'motifs' that are recognizable when we compare many factors. A list of motifs has been established: zinc fingers, leucine zippers, steroid receptors, helix-turn-

The organizer 153

helix, and amphipatic helix-loop-helix; the last two motifs are defined according to their form, the others display specific chemicals.

One of these motifs is particularly interesting. The helix-turn-helix motif consists of two helicoidal branches, one that lies in a wide DNA channel, and another that lies at an angle across it. And, interestingly enough, a related form of this motif may be present in the homeobox of *Drosophila* and of several mammals (Tamkun & alia, 1992, among many other papers). This means that the group of genes that contain homeobox sequences presumably intervenes in both the activation of transcription, and in the specification of body parts during morphogenesis. Processes inside and outside the cell are not only analogical processes, as close as the analogy might be: in the case we are considering they are much more, they are similar processes and, consequently, they are related to similar genes.

We may then conclude in relation to homeobox that the tissue level has not entirely emerged from cell level in morphogenesis as a new complex system, with its own rules and units, but rather that it also continues the cell level because the former benefits from a mechanism that is operative in both. No wonder homeobox-gene clusters have maintained essentially unchanged through evolution: they look as stable as the genetic code itself. However, the most interesting conclusion that may be drawn is not that the product works just inside the cell, merely decoding the DNA, but it is also able to work outside in the intercellular space. Synthetases share this property also: they charge tRNA with specific amino acids and at the same time they intervene in the embryo's growth. The circumstance I wish to emphasize here is that the plural value of an enzyme extends over its form: it is not only that the same protein participates in transcription and in morphogenesis, it is that some of its formal patterns, the so called motifs, do it at the same time. Briefly: *organisms seem to have inherited a formal pattern*, just the methodological heresy whose possibility we were obliged to reject in chapter VII.

10.6 Genetic code meets linguistic code

Here, we reconsider the hypothesis behind the argument of this book. The verb *to inform* (*somebody about something*) habitually means *to bring up to date*; but it has another meaning: *to give a special quality or character*, that is, *to give form*. The first one *informs* from the outside, the second one *in-forms* inside. So far we have compared genetic code with linguistic code and we have concluded that both of them are informational devices. In the first sense of the root *inform*, genetic code, or, rather, its manifestation by means of a particular genome, allows organisms to transmit from every individual a strategy that monitors the reaction of the organism to the changes of the environment to its descendant. Similarly, the linguistic code, that humans may have inherited to some extent from their ancestors, allows a speaker to transmit to the listener any state of affairs that affects them, whether a real or a mental state, by means of a given message.

But genetic code and linguistic code also share the second meaning of the root *inform*. Genetic code, which is manifested by means of a particular genome, gives form to the organism that has inherited it, and is being responsible for the particular pattern that differentiates every species and even every individual from one another. Linguistic code does the same: humans acquire the biological form that characterizes them from all the other animals, and the most striking feature they display in front of them is precisely their ability to speak a language, a kind of behaviour that only linguistic code makes possible.

Erwin Schrödinger pointed this out in his well known essay on the origins of life (Schrödinger, 1944) that the chromosome structures are at the same time instrumental in bringing about the development they foreshadow. And this also happens to be the case with linguistic structures. Language is a formal device that informs us of the state of the world and that in-forms the world it lives in turn. Children that fail to acquire a human language when growing up suffer two impairments: they do not receive useful data of the environment, and, most importantly, they interrupt their own developmental process. Should we still persist in thinking that formal parallelism of genetic code and of linguistic code to be originated at random?

Growing up poses a serious challenge to the embryos of any animal species: they react adequately to it by following the instructions coded in their genome. But human growth is much more difficult than the growth of other animals: infants are born prematurely, hence they have to acquire from the outside many resources they could not obtain before birth. Most of them constitute a learned resort, quite sure, for humans are the species that possess the greatest amount of cognitive capacities, a set of devices that confers on them self-awareness, technological tools that permits them to change their environment all over the world, an so on. But language, although mostly gained through learning, cannot be entirely acquired this

The organizer

way, as we have already mentioned. Thus, it has to rely, to some extent, on the genome.

Unfortunately we have examined the reasons why a strictly hereditary process for language faculty seems implausible: neither a considerable mutation process, nor a weakened versions of it, such as the Baldwin effect or exaptation, moved to fill the gap. There is, however, an alternative solution. The surprising inheritance of formal motifs, from the cell nucleus to the morphogenetic processes that are regulated by homeobox genes, leads us to hazard a guess at this puzzling problem: it could be that the genetic code had formally influenced the structure of linguistic code, which not only shares some structural features, but also works in a similar way. In other words: *genetic codes could have been the pre-program linguistic codes came from.*

11 The limits of complexity

11.1 A double source for complexity

A very common difficulty that arises when considering the emergence of complex self-organizing systems is the degree of complexity those patterns have to attain before it can be said that a new level has been born. The problem of the origin of life is a good example. Since Stanley Miller succeeded in performing his famous experiment fifty years ago, there is no doubt that life on earth was born out of some chemical reactions that took place among a limited set of molecules (H_2O, CO_2, NH_3, CH_4...), which were made out of H, O, N, S, P, and especially C. But after having satisfactorily explained the emergence of amino acids, there's still a long way to go. First of all, although complexity is a necessary condition for life it is not in itself sufficient: there are many complex molecules that include more than the twenty amino acids, and which form many proteins that cannot be produced by the living beings. Secondly, and most importantly: sometimes complexity does not consist of the emergence of new properties, but rather of the cooperative association of phenomena that were unrelated to each other at the previous level.

The first type of difficulty is usually overcome by finding a property that exclusively characterizes the complex units we are interested in. This was the case with the amino acids of life: from a chemical point of view they do not constitute a special set despite the fact that all of them are related to a DNA triplet according to the correspondences established by the genetic code. In other words, the problem of amino acids was converted into the problem of DNA strings. As a result of this, a different kind of complex structure becomes the central point of genetic research. From now on, the problem is no longer the length of polypeptidic chains, and the secondary topological structures they adopt when bending, but the way the double helix of DNA is constituted.

Notice, however, that by converting the problem of amino acids into the problem of DNA, we change the perspective of the research: the most intriguing fact no longer remains in the chemical nature of nucleic acids, nor in the reason why Adenine claims for Thymine, and Cytosine claims for Guanine, but in a structural property that only DNA and RNA share: the ability to make copies of themselves. Hence, much attention has recently been paid to ribozymes. S. Kauffman (2000, ch. 2) points out that the weakness of the classical model as conceived by Watson and Crick lies in the impossibility of DNA to replicate and translate into amino acids without the help of several enzymes, that is, without some proteins, which had to be coded by DNA itself. A modern cell is an autocatalitic collective whole where DNA, RNA, the code, the proteins, and the metabolism cooperatively act together. Unfortunately, nobody has succeeded until now in replicating a single chain of DNA or RNA in the laboratory. No wonder if we consider that the problem researchers had to solve was a very difficult one.

Unless an organism capable of replicating its DNA or RNA without proteins be found. *Tetrahymena*, a unicellular being that was discovered a decade ago, seemed to be this organism. Scholars found evidence that *Tetrahymena* has DNA introns, which code for RNA introns. These introns are autocatalitic: before the RNA string is translated into a protein, their RNA introns are spliced out and the remaining exons are joined together in a pathway that is catalysed by the intron itself. Now we know that all eukaryotes possess introns. We may suppose, then, that the first systems contained only a self-replicating nucleic acid with primitive catalytic activities. Proteins had been added later, as a stabilizing device that maintains precarious RNA alive. A long step further on evolution produced that proteins took over catalytic reactions while RNA was relegated to a structural function.

As complicated as this process can appear, the theory of complexity accounts for it in a very elegant way. Kauffman (2000) suggests that life did not begin as a replication of moulds, but rather as a catalitic closure. Experiments have been done where some nucleic tetramers catalyse two dimers obtaining a second tetramer, which is exactly similar to the first one. In other words, A joins two parts of A together in order to get A. But experiments by Gunter von Kiedrowski went a step further: we can also see that a molecule A catalyses the sticking of parts of B and, vice versa, that B catalyses pieces of A that are joined together. Kauffman has formalized the emergence of complex biological structures by means of a mathematical model: the so-called reactive graph. The idea is as follows: in a set of molecular species that is either a substratum, or a product, or both, and that are capable of catalysing the formation of other molecules, phase transitions necessarily appear after a certain number of connexions are made. These phase transitions produce the emergence of organic systems: evolution and the birth or new species are supported by such an approach.

The explanation of the origin of life by means of the theory of complexity seems to leave no loose ends. In fact, neo-Darwinian theory of evolution by variation and natural selection is nicely complemented by the theory of complexity. But there is still one remaining question. We have said that prokaryotes like *Tetrahymena* possess autocatalytic RNA introns. However, replication, transcription and translation, the process that allows eukaryotes to disseminate copies of their genome in their descendants and – in pluricellular organisms – to any cell of their own body, is a very complex set of operations that can hardly be related to the autocatalytic prokaryote by single gradual evolutive pathways. One wonders, therefore, how could prokaryotes, which already existed a thousand million years (3,500 millions) after the Big Bang (4,500 millions), evolve until they were converted into eukaryotes two thousand million years later (1,500 millions), no intermediate organisms permitting us to follow their tracks. In other words: we are able to account for the increase of complexity from the individual eukaryotic cell until the human brain, the most complex object of nature, but we do not know how to explain the first step, the emergence of complexity patterns in the spliceosome, the factory of replication, of transcription, and of translation.

A theory that challenges the neo-Darwinian model of the synthesis helps us to overcome this drawback: the symbiotic model by Lynn Margulis (1998). Margulis supposes that eukaryotes are the result of joining the genomes of two (and perhaps more) prokaryotes when one of them incorporated – that is, swallowed – the other. This is obvious in the case of mitochondria and of chloroplasts, which still have a slightly different DNA from the DNA of the nucleus, but Margulis extends her hypothesis to the nucleus itself whose DNA would be a mixed one, some genes coming from a gram-positive bacteria and some others from a gram-negative bacteria. Once the eukaryotic level had been attained, evolution continued unstoppably by combining natural selection and the emergence of complex structures.

We may conclude that the origin of life and the formation of living organisms is a process that has developed in three steps:
i) The gradual (i.e. Darwinian) emergence of a complex new auto-organizing level: the prokaryote cell (bacteria and other similar organisms);
ii) The sudden appearance of a symbiotic unit that combines some structural properties of organisms of the lower level: the eukaryotic cell (amoeba, etc.);
iii) The gradual (i.e. Darwinian) emergence of successively more complex species, from the unicellular eukaryotes until the mammals (humans among them).

11.2 The origin of language: a gradual process with a break

Life was the result of a process that developed according to the following curve:

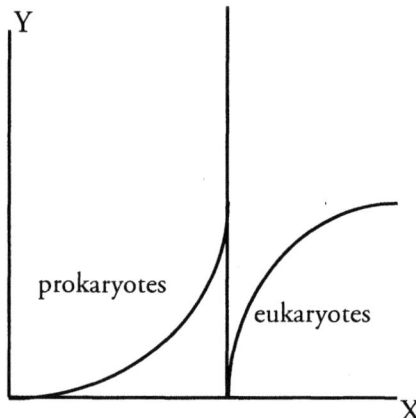

where the section on the left represents the progressive constitution of autocatalytic sets of RNA molecules that characterize viruses – the first living beings – and the posterior prokaryotes with DNA and RNA, whereas the section on the right represents the world of eukaryotic organisms from unicellular beings until to humans. Between every species and the following one there can be a smooth transition – as stated by orthodox neo-Darwinism –, or there can be a more sudden transition – according to the punctuated equilibrium approach –. In both cases, however, no radical gap separates any organism from its immediate descendant. But the pathway that goes from prokaryotes to eukaryotes is much more than a transition. From a biological point of view, the line that separates prokaryotic and eukaryotic cells is drawn by the most important categorial criteria in the story of life. No wonder that life continued basically unchanged for 2,000 million years since it appeared on earth 3,500 million years ago, while the eukaryotic world passed from amoeba to humans in only 1,500 million years. The common idea that a bacteria, like *Escherichia coli*, and a protist, like an amoeba, are very similar "bugs", when the last one differs greatly from a fly, from a frog, and even essentially from a human, is wrong. Perhaps people convinced themselves of this idea because they can "see" eukaryotic organisms bigger than a flea, but unable to see anything smaller without a microscope. Anyway, the most important gap in the story of life opposes the

The limits of complexity

organisms with DNA, but without nucleus, to the organisms that possess a nucleus, which includes the chromosomes, and a cytoplasm.

Turning now to the problem of the origin of language, we are faced with a very similar situation. It is reasonable to suppose that perceptual abilities slowly evolved from amoeba to big apes and hominids, who were already capable of representing the external world in their minds by means of a sophisticated visual system and probably also by means of an incipient protolanguage. But humans are not hominids: it seems that *Homo habilis*, according to palaeontological remains relative to the form of jaw, the position of larynx, and the size of brain, was already capable of speaking. From then, that is, two million years ago, language converted itself in the very complex system we know in *Homo sapiens*, our own species, which appeared on earth 100,000 years ago. As in the case of the origin of life, the process that gave rise to language distinguishes, thus, two gradual phases that are separated by a gap (perhaps this gap is situated between *Homo habilis* and *Homo sapiens*, rather than between *Australopithecus* and *Homo habilis*, but this would be a secondary issue):

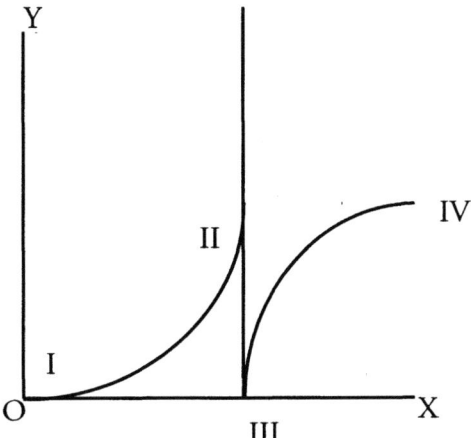

We recognize the following phases in the curve above, where the axis OY stands for complexity, and the axis OX stands for time:

From point I until point II eukaryotic organisms gradually improve their perceptual apparatus. In I, amoeba and other protists simply react to alterations in the environment by stopping the movement of their cytoplasm (pseudopodon). As intermediate steps we may select the progressive development of the eye (Bruce and Green, 1990), and of the other perceptual organs. For example, molluscs, like mussels, have a series of ocular spots alongside their shell, which allow them to notice a predator that moves over them because the intensity of light changes. Insects, like bees, in their turn, have built a complex eye by joining a modified version of these ocular spots together. Amphibians, like frogs, have already got an eye that is very similar to ours: however, their brain still works a bit crudely, it does not distinguish different objects, except only either big moving surfaces (predators), or small moving ones (preys). Mammals, apes among them, are supplied with a very sophisticated visual system, which not only recognizes surfaces that stand up, but also the kind of object they represent. When this section of the curve approaches point II, a protolanguage appears: apes and hominids that have substituted the visual input for a phonetic one, and that have weakened the iconic relationship of the symbol to its referent, get an incipient communicative tool, which underlies a society of several dozen of individuals where sexual exchanges are of great importance.

The step that goes from III to IV has been carefully investigated by Wildgen (2004) who emphasizes that the integration of the evolutionary dimension in linguistics represents a fundamental turning point in the history of this discipline. Such a trend should expand Grimm's time span of 2,000 years (when the first consonantal change, that separated the Germanic languages from their Indo-European stem, the so called Grimm's law, took place) up to 100,000 years, and we need to consider some semiotic aspects far removed from the traditional concerns of linguistic theory. Wildgen has shown that there are some fundamental action schemata, which are based on motor control by hands in the design, manipulation and use of tools, as palaeontological and archaeological testimonials clearly manifest: these schemata belong to a cognitive-semiotic layer that other animals never attained. Wildgen points out that these schemata form a hierarchy of dynamic archetypes that define evolutionary thresholds, which had to be overcome. Every level of this evolutionary process (the step III > IV in my own model) represents an equilibrium between the assessment of a cognitive level and the cultural communicative profit derived from it. The transitions from each level to the next more complex one is considered to be rather smooth and not catastrophic, although the thresholds exist.

I agree with Wildgen except for his idea that the syntactic formal properties that Chomsky calls the I-language, a kind of universal grammar, are epiphenomenic features, which appear rather late in evolution. If they were epiphenomenic, languages would differ with respect to these syntactic properties, in the same way as they have developed different phonetic or lexical systems. Perhaps most generative grammar insights are wrong and actually

I think some of them are: language building is much more related to the perceptual circumstances that allow it to re-present the external world than to unrooted formal constraints. However some abstract syntactic properties appear time and again. It seems quite difficult to ignore the so called Plato's problem, the fact that children suddenly improve their linguistic learning process – no matter the natural language they are acquiring – at the age of two. This always happens in approximately the same manner, irrespective of the quality of the input they receive, of their level of intelligence, or of the help they get from the adults. And most important of all: this happens when the protolanguage, the baby talk they have previously mastered, is converted into a language because those formal syntactic properties emerge.

Moreover, authors that are used to emphasizing the semiotic aspect of language should be aware that this formal syntactic framework represents a necessary condition for linguistic cultures to emerge. Brandt (1994) points out that what distinguishes a linguistic text from any other text – as found in a computer program or in a similar system that processes information – is the fact that the linguistic text consists of modalized actantial schemata which convert somatic acts into discursive acts. In other words, actantial schemata are not discursive, actantial schemata arise among animals as soon as they interact. Humans also interact with each others, and thus develop somatic action schemata like any other animal. But humans can go a step further and communicate those actantial schemata by means of another action scheme (I say that X to you). This communicative status is only attained when the multivariable actantial schemata of the perceived world pass through two new levels of formalization:

First, when an actant is opposed, as a subject, to the remaining actants which constitute the predicate, and, eventually, local adverbs are inserted that settle a deictic centre for the listener;
Second, when the speaker introduces a new point of view by opposing a nominal or adverbial element, as theme, to the rest which is the rheme, and, eventually, by adding modal and temporal values with auxiliary verbs or with adverbs.

Linguistic semiotics is the study of the procedures by which linguistic texts are modalized and deictically anchored. However, it is worth noticing that these functions, modality and deixis, cannot emerge without a previous formal background. They do not emerge automatically from complex actantial schemata because they have not come into being in the very complex societies of social insects or primates, which preceded humans on earth.

Predication and thematisation do not appear as a consequence of increasing the number of actants and actantial schemata, they appear as a new semiotic dimension, an intensive one, which enables the force conflicts that spring up at the extensive layer to be resolved. The formal properties we have considered above just underlie the predicative linguistic behaviour (categories, agreement, etc.) and the thematic linguistic behaviour (anaphora, focus, etc.): this is a remarkable fact semioticians have to challenge.

Unfortunately, the drawback generative grammar – the school that primarily described these formal properties – was faced with has been always the same: firstly, such a sudden emergence suggests a kind of catastrophic process, say a big mutation, which the state of art of current biological knowledge absolutely excludes; and secondly, the specific syntactic properties we are speaking of lack biological justification. I think we must agree with the editors of a recent readings book on the topic (Knight, Studdert-Kennedy and Hurford, 2000, 1) when they point out:

> As a feature of life on earth, language is one of science's great remaining mysteries. A central difficulty is that it appears so radically incommensurate with non-human systems of communication as to cast doubt on standard neo-Darwinian accounts of its evolution by natural selection. Yet scientific (as opposed to religious or philosophical) arguments for a discontinuity between human and animal communication have come into prominence only over the past 40 years.

Chomsky proposed a distinction between mysteries and problems in science: according to his proposal, the emergence of language is 'mysterious'. But the neo-Darwinian paradigm has come to cast some light on the subject simply by developing the idea that language is no ordinary adaptation (Maynard Smith and Szathmáry, 1995). The explanation I am proposing here is aimed to fill the gap between I > II and III > IV assuming something that I consider obvious, that is, the fact that there is a gap between II and III.

11.3 Syntax as a symbiosis of several genetic encodages

My proposal will satisfy, I hope, both the above requirements at the same time: it does not require a big mutation but a symbiotic process which benefits from previously existing biological entities. Scholars usually agree on the challenge the Chomskian paradigm represents to Biology. But the most difficult fact is not the epistemological implausibility of the sudden emergence of syntax, but the idea Chomsky has of syntax itself. Let us reproduce his own words relative to the minimalist approach (Chomsky, 1996, 171):

> Nevertheless, it seems that economy principles of the kind explored in early work play a significant role in accounting for properties of language. With a proper formulation of such principles, it may be possible to move toward the minimalist design: a theory of language that takes a linguistic expression to be nothing other than a formal object that satisfies the interface conditions in the optimal way [italics mine]. A still further step would be to show that the basic principles of language are formulated in terms of notions drawn from the domain of (virtual) conceptual necessity.

Here Chomsky slips into a contradiction. Unfortunately, his optimality theory cannot have a biological foundation anyway because optimal solutions are not the solutions of natural selection. Jacob (1977) pointed out that evolution does not work as an engineer, but as a *bricoleur*. This means that there is nothing like a perfect design every time a species converts into a new one. Engineers design the system they want according to a previously concerned formal plan. Nature, on the contrary, simply chooses from among the several possibilities genetic variation put at its disposal. The selected one will be good, sure, but not optimal simply because it was attained after and not before genetic variation. It is the best solution, the one that adapts most easily to the environment, but no one could guarantee that it responds to a conceptual necessity. Natural selection is a bricoleur that picks up several objects, which are randomly joined in the shed in order to make a tool that permits you to react to the challenges of the external world. People who have a back pain know what I mean: most times it is a consequence of the standing up posture our ancestors acquired through natural selection several millions of years ago.

The hypothesis I am formulating here is strictly Chomskian in the sense that it accepts the principles and parameters properties of syntax as the starting point of the emergence of human language, but it does not accept his explanation that they are what we know because they are optimal solutions. They are simply some properties of our genetic endowment, which natural selection benefited from in order to constitute a powerful tool that our predecessors (no matter if whether they were *Australopithecus, Homo habilis* or even *Homo erectus*) could never have dreamed of.

Let us turn to the problem of the origin of life, and particularly to the appearance of eukaryotic cells, the first organisms to bear information and which evolution was slowly improving until one of their descendants began to speak. The symbiotic process described by Margulis consists of three main components:

1) A Gram-positive bacteria, which furnished the coding component, the genes responsible for replication of DNA, transcription into RNA, and translation into proteins;
2) A Gram-negative bacteria, which contributed with the patterning component, the genes that give rise to microtubules and membranes that form an extensive network through the cytoplasm;
3) A third bacteria, cianobacteria in plants, that gives rise to the energetic component, the organelles like mitochondrion and chloroplast (in plant cells) that supply the cell with energy by getting it from the outside.

The symbiotic process that took place 1,500 millions of years ago did not simply join together three components whatsoever. It constructed a tool with a coding, a patterning, and an energetic component at the same time. This is not surprising: living organisms are just those beings that are capable of passing on their identity to the descendants whilst remaining isolated from the environment, and benefiting from the energy provided by the external surroundings at the same time. The eukaryotic cell, which came out from three previous prokaryotic cells, was not optimal, but it represented the best solution under the circumstances into which it was born.

It stands to reason that language is a powerful tool that provided humans with an adaptive advantage over their competitors. This tool consists necessarily of three formal components:

1) A set of categories and rules that the speaker and the listener share in common so that the latter can understand the former: this is the *coding component* of language;
2) A pattern that holds the elements of the message joined together: this is the *patterning component* of language;
3) A flow of information that transmits the intentional meaning from the speaker to the listener: this is the *energetic component* of language.

Now consider the problem our predecessors were faced with. The size of social groups was growing, thus allowing them to react much more adequately to the challenges of the environment. But social complexity had arrived to a communicative barrier. Protolanguage was no longer capable of representing the sophisticated images of the mind that were required by growing social interaction. Neural cells, the cells that supported these mental processes, and are hard wired in the brain, reacted to the challenge by combining the genetic instructions relative to three different aspects of the problem:

1) Genes responsible for the behaviour of the code, that is, those genes that put the correct nucleotide base (A, T, C or G) in complementary base pairing, during replication, and those genes responsible for translation, that is, for putting the right amino acid in relation to each nucleic base triplet, constituted a subsystem, the coding component or C-subsystem.
2) Genes responsible for the behaviour of non-coding satellite DNA, constituted another subsystem, the patterning component or P-subsystem.
3) Genes responsible for controlling genetic transcription, that is, regulator and operator genes, constituted a third subsystem, the energetic component or E-subsystem.

Was it a catastrophic change? Not at all. Notice that the opposition between structural and regulator genes belongs to the nature of the genome itself: The only innovative change is the differentiation of structural activity proper (that is, C-genes) and satellite structural activity (that is, P-genes). However, the human genome is characterized within animal genomes for its satellite DNA – and especially the so called Alu sequences – continuously growing and occupying a major extension of the chromosome.

It is true, that – apparently – happened when life appeared on earth, and the process I am hypothesizing here differs in one important aspect. The first consists of three organisms (three prokaryotes) that fuse together to form a new organism (an eukaryotic cell): this was an external process. By contrast, the second one would rather be an internal process where the neural cells of a human brain experienced a severe rewiring of their connexions which resulted in a threefold genetically determined system constituted by C-subsystem, P-subsystem, and E-subsystem. This is the way a symbolic preprogram could have developed. Since formal structures and processes from inside the cell can sometimes be maintained and operating outside it, as we have already seen (cfr. § 9.4), a similar extension could have happened in this case, unfortunately we lack experimental evidence of this, and so it must remain only a hypothesis.

A suitable hypothesis in any case, I think. Its biological achievement does not seem impossible. The only two conditions that have to be fulfilled are, firstly, a source of additional DNA where experiments can be realized without affecting the rest of the genome, and, secondly, a mechanism that allows this extra DNA strings to put the genes responsible for the C-subsystem, for the P-subsystem and for the E-subsystem together. The development of germ cells provides a suitable model for the first condition. Eggs vary enormously in size between different animals, and they are much bigger than their corresponding somatic cells in any case. There are several procedures for oocyte development, either gene amplification, or contributions from other cells. The strategy vertebrates usually adopt is to increase the overall number of gene copies in the development: this proportionately increases the amount of mRNA that can be transcribed, and thus the amount of protein this mRNA codes for. Maybe, even though some of these extra copies of the genome did not convert into protein, they provided the basis for a gene rearrangement. Two general procedures may be hypothesized in order to explain this rearrangement. One of them is splicing and posterior mixing up of the exons: RNA splicing involves breaking of exon-intron boundaries and forming a bond between the end of the exons after removing of the introns; sometimes free exons are mixed up while splicing and a new arrangement is obtained. Another procedure is transposition, which would affect DNA and not RNA: as above, duplication of sequences within the genome would be the previous step; whereas one copy can retain its original function, the other may evolve into a new one because some of their elements are transposed.

The difference between the first procedure – splicing – and the second procedure – transposition – lies in the fact that RNA is ephemeral, but DNA is not. In other words: to join C-, P-, and E-subsystems together by splicing and subsequent mixing up of the exons (but repetitive DNA being lost!) would give rise to an occasional synaptic web which supported the grammatical properties of language in a given communicative occurrence. The human being that casually produced such a grammatical sentence for the first time would simply have retained these procedures by hearth and culture would lately disseminated them, as a set of memes, through society. This explanation reconciles, then, the surprising parallelisms between linguistic code and genetic code with any behavioural theory of the origin of language such as Darwin's. On the contrary, if C-, P-, and E-subsystems were put together by means of transposition of fragments of DNA, then they were incorporated into the genome of the first human speakers and the grammatical procedures were simply inherited by their descendants. I rather support this last possibility because the complexity of these grammatical procedures makes it difficult to acquire them all at once. However, the genetic-like character of the linguistic code that was already emphasized by R. Jakobson is independent of this point of view, whether we adopt innate or acquired, to explain the origin of grammar: the first one seems simply more acceptable.

12 Some concluding remarks

We turn to the parallelism pointed out by Jakobson (1971) again. Scholars habitually ignore it or at most they are used to minimizing its importance. A way the similarity of the linguistic and of the genetic code is played down has been to retain only the formal aspect of their proximity. For example, Abler (1989) extended Fisher's (1930) arguments concerning biological inheritance to chemistry and language. Fisher pointed out that variation, the mechanism natural selection benefits from, requires that the characteristics of the parents do not blend, but that they can reappear in the descendants. These characteristics are not lost because the mechanism of biological evolution depends on a set of *particulate factors*, the genes, as Mendel demonstrated conclusively. The diversity of individuals of a species – and finally the diversity of species itself – is a consequence of this fact: the properties of one of them do not lie between those of their parents, but outside, which allows variation to be conserved or even increased across generations.

Abler (1989) noticed that the particulate principle mechanism works also in chemistry and in language. Hundred chemical elements, which Mendeleiev collected in his well known periodic table, combine to generate the infinite number of compounds the natural world is made out of. The great variation we get this way follows the particulate principle law: the characteristics of the atoms that join together do not reappear in the resulting molecule, for example, C burns and O helps burning, but CO_2 is not inflammable. Similarly, as Wilhelm von Humboldt noticed already in 1836, language makes an infinite use of finite procedures: the set of morphemes that constitute the lexical inventory of any natural language is poorly mastered by their speakers, which are used to knowing only a few thousands, but it allows them to construct an infinite number of sentences. Moreover these sentences exhibit many properties that were not present at the compounding morphemes: for example, we cannot predict that *came* will mean a state until we join it with *true* (*my dreams came true*), we cannot predict that it will mean a process until it meets *decision* (*we came to a decision*), and we cannot predict that it will mean an action until we combine it with *village* (*they came to a village*).

Unfortunately these remarks went a step further in Abler (1997) when he relates the particulate principle of chemistry, genetics and language to numbers too. On the one hand, genes are DNA chains made out of chemical molecules: no wonder that they follow the same principle of organization to some extent. On the other hand, there is not such a straight link between the genome and language: however both are organized in texts, and, thus, it seems conceivable that the former ones have influenced the latter ones in some way. But when numbers are added to the three domains above, empirical issues are substituted by formal ones. According to Abler's view, the particulate principle would be a formal property of the physical world, a property that numbers exhibit on their own and which is manifest in the form of molecules, genomes, or linguistic texts among many others.

No doubt he is right: complexity requires infinite products (be they molecules, genes, linguistic texts, or numbers) to be obtained by means of finite resources. But the question remains as to whether this is all we can say about the problem. We intended to prove that in the case of the relationship of genetics to linguistics (as in the chemical-genetic connexion) there is much more to be considered. There are too many parallelisms for a simple formal relationship to underlie them. Neither chemical compounds nor the combinations of numbers look like the groups of linguistic units. As I have shown, only the behaviour of the genome closely reminds us of the behaviour of linguistic texts. And this, no matter whether the particulate principle holds true, is the problem that has to be explained if we are to understand how language could have originated.

My own position in relation to the Chomskian hypothesis is ambiguous. I and many other scholars have tried to reduce the extension of his universal grammar by showing that most properties of human languages are the result of the perceptual grasping of the world these very languages allow. The amount of assumed universal features of human language should be drasticalley reduced, then, according to the research done over the last twenty years. Chomsky reluctantly agreed: since *Syntactic Structures* (1957) those alleged innate syntactic properties decreased every time a new proposal was made by him. However, now I believe Chomsky gave in to his opponents too much in his last *The Minimalist Program* (1995). This claim has been sustained from inside the generative grammar as well as from outside. For example, F. Newmeyer (1998, 317–318), a well known grammarian who contributed to spreading the Chomskian paradigm worldwide, wrote:

> Chomsky no longer conceives his unexplicated physical principles to have engendered all of the messy parameterized and often complex principles that were once attributed to UG – only their residue subsumed by the economy principles. But it still leaves us with the problem of accounting for the origins of the vast amount of grammatical properties that are still arguably provided by an innate UG, yet in no way reflect economy principles [...] Noam Chomsky's ideas have, quite properly, set the research agenda for linguistic theory for several decades. Such ideas lead inexorably to the postulation of genetically determined aspects of grammar, which, in turn, naturally invite inquiry as to their phylogenesis. Unfortunately, Chomsky has either disparaged the value of such inquiry or advanced ideas that seem, at one and the same time, generally inadequate and mutually contradictory.

And Li & Hombert (2002, note 14) from the functionalist (i.e. adaptionist) side point out on their turn:

> We take note of the fact that Chomsky's current theoretical stance is considerably different from his 1986 pronouncements. In his new Minimalist Program (Chomsky, 1995), grammar is largely derived from the lexicon. If we are correct in assuming that what is considered innate by Chomsky and his followers is the newest version of the so-called «Universal Grammar», which is austere and minimal, the issue of representational innateness for language behavior is practically moot.

The story reminds me of a similar story regarding Einstein. His theory of relativity laid the foundations for an explanation of the origin of the universe. However, Einstein did not accept the Big Bang hypothesis, and he added the cosmological constant to his formula to support the so-called Constant State Physics. Only after many years, was latter empirical evidence of the Big Bang (ground radiation) discovered, Einstein changed his mind in a direction that meets his own theory. This could also be the case with Chomsky. His intellectual trajectory can be characterized as an attempt to demonstrate that there are some grammatical properties, which are not motivated (that is, contextual free ones), and which all languages share. These properties, which constitute the universal grammar (UG), should be innate, and generative grammarians progressively made their inventory smaller and more accurate from 1957 to 1981 (the principles and parameters model). Many scholars – I among them – have been reluctant to accept the Chomskian paradigm because we think linguistics is much more than the reductionist view Chomsky provided us with. In particular, we think that contextual dependence cannot be laid aside when considering linguistic facts. Anyway, the steps attained by Chomsky are real steps: the principles of the 1981 paradigm seem to be universal and unmotivated.

But such an impressive, though reductive, hypothesis clashed with the Darwinian paradigm. The properties of UG must be innate. And if they are innate, they must have been acquired through natural selection. Chomsky was aware of the difficulty that his theory represented for Darwinism because the UG properties are formal features that do not contribute to improving fitness at all. The answer to the challenge was: there is a problem, quite certainly, but biologists are facing up to it, it is not the linguists' task. Unfortunately, it seems biologists are not very keen on linguistic problems. Actually, they did not solve the problem Chomsky had pointed out. Hence, it was Chomsky himself who was committed to explaining how language could have arisen in a hominid's brain. The strategy he adopted was the minimalist program (1995): let us simplify the amount and formal complexity of the properties humans are supposed to share, until we come to a greatly reduced inventory, which could have emerged from the relations several lexical items maintained in the mind.

This is the current proposal of generative grammar, a proposal that is supposedly compatible with Darwinism, although it constitutes a poor basis for the understanding of grammar. Moreover, these MP linguistic properties are not only poor, they are shared with any set of discrete units capable of infinitely combining with each other, be they chemical elements, or numbers. However, surprisingly enough, the linguistic principles of the P&P paradigm supposedly emerge from MP, but numerical or chemical MP did not give rise to numerical or chemical P&P principles. It looks as if Chomsky had ballasted his theory with a heavy counterweight. Like Einstein did.

There is another path I have tried to develop in this book, although some scholars have also made valuable suggestions that helped to open it (Collado-Vides, 1996, Raible, 2003: see López-García, 2005, for discussion). Instead of simplifying the linguistic laws in order to accommodate them to the biological reality, we could, instead, look deeply into the very biological reality until we are able to find some characteristics that accommodate the linguistic laws. In other words: we must find the ground radiation that underlies the Big Bang and the theory of relativity at the same time. This renewed vision consists of several steps:

Firstly, there is an extended parallelism between the genetic code and the linguistic code;
Secondly, biological forms like homeobox may be inherited;
Thirdly, there is also a parallelism between some intracellular motifs and some extracellular forms, which developed from them;

Fourthly, the genetic code manifests as a set of intracellular formal procedures; linguistic code manifests as a set of brain wired extracellular connexions;

Fifthly, the genetic code could have given rise to the linguistic code (to the P&P laws) by genome duplication and the subsequent inheritance of formal structures. As reported in *Nature* 431 (21 october 2004), the total number of protein-coding genes of the human genome is in the range 20,000–25,000, a number which resembles much closer those of worms than those of mice or rats. Nevertheless, the proportion of segmental duplication is clearly higher in the human genome than in the mouse or rat genomes. Duplications arising after divergence from the rodent lineage affect a total of 1,183 genes, which are presumably responsible for human specificity.

Such an explanation, which supposes the existence of a break in the middle of a gradual process, is compatible with both equilibrium states that precede and follow it. The former is the slow enhancing of the perceptual mechanisms of animals up until the sophisticated visual apparatus of mammals (and lastly the protolanguage with its MP laws); the latter is the spreading out of the formal P&P laws in a set of languages, each of them representing an accommodation of these laws to the perception of the external world. Moreover, such an explanation accommodates neither formal linguistics (generative grammar) only, nor functional linguistics (semiotics and cognitive grammar) only: as a matter of fact, the formal laws of syntax are certainly innate but because they resemble the genetic code, that is, because they are an instance of embodiment.

Since the Homo floresiensis was reported some months ago no direct correspondence between mental development and brain size can be established. Those people had only a 500 c.c. brain capacity (like a chimpanzee) but the cultural artefacts that have been found around them suggest they were some kind of speaking species. This challenges current assumptions on the iconic origin of verbal behavior – embodiment – for the external world has mapped into very different neural webs in essentially the same way. There must exist some kind of formal pattern that underlies all human languages irrespective of the surrounding conditions where each one grew up. The hypothesis of this book is that DNA formal laws have been, through genome duplication, the structural model followed by neural connexions every time a group of hominids developed a language and converted into humans. Genomic laws filtered the iconic process, and humans – opposite to animals – experienced a twofold embodiment.

References

ABLER, W. (1989), "On the particulate principle of self-diversifying systems", *Journal of Social and Biological Structures*, 12, 1–13.
ABLER, W. (1997), "Gene, language, number; the particulate principle in nature", *Evolutionary Theory*, 11, 237–248.
AIELLO, L. C. and DUNBAR, R. I. M. (1993), "Neocortex size, group size and the evolution of language", *Current Anthropology*, 34, 184–195.
ANDERSON, S. R. and LIGHTFOOT, D. W. (2002), *The Language Organ: Linguistics as Cognitive Physiology*, Cambridge University Press.
ARMSTRONG, D., STOKOE, W. and WILCOX, S. (1995), *Gesture and the Nature of Language*, Cambridge, Cambridge University Press.
BAKER, M. (1997), "Thematic roles and syntactic structure", in L. Haegeman, *Elements of grammar. Handbook of Generative Syntax*, Dordrecht, Kluiver, 73–137.
BALLY, Ch. (1932), *Linguistique générale et linguistique française*, Berne, Francke.
BARNDEN, J. (1998), "An AI System for Metaphorical Reasoning about Mental States in Discourse", in J. P. Koenig (ed.), *Discourse and Cognition*, Stanford, CSLI, 167–188.
BERGOUNIOUX, G. (2002), "La sélection des langues: darwinisme et linguistique", *Langages*, 146, 7–16.
BERWICK, R. C. (1998), "Language evolution and the Minimalist Program: the origins of syntax", in J. R. Hurford, M. Studdert-Kennedy and Ch. Knight (eds.), *Approaches to the Evolution of Language*, Cambridge, Cambridsge University Press, 320–341.
BICKERTON, D. (1990), *Language and Species*, Chicago, Chicago University Press.
BICKERTON, D. (1998), "Catastrophic evolution: the case for a single step from proto-language to full human language", in J. R. Hurford, M. Studdert-Kennedy, and C. Knight (eds.), *Approaches to the Evolution of Language: Social and Cognitive Bases*, Cambridge University Press, 341–358.
BICKERTON, D. (2000), "Calls aren't words, syllables aren't syntax", *Psychology*, 11, 114.
BLOOMFIELD, L. (1933), *Language*, The University of Chicago Press.
BOUWMEESTER, T. (2001), "The Spemann-Mangold organizer: the control of fate specification and morphogenetic rearrangements during gastrulation in *Xenopus*", *International Journal of Developmental Biology*, 45, 251–258.
BRAND, P. A. (1994), *Dynamiques du sens*, Aarhus University Press.
BRAND, P. A. (2002), "Music and the abstract brain", *The First International Conference on Neuroesthetics*, University of California at Berkeley.
BRUCE, V. and GREEN, P. R. (1990), *Visual Perception. Physiology, Psychology, and Ecology*, London, Lawrence Erlbaum.
BUCK, I. and AXEL, R. (1991), "A Novel Multi-Gene Family May Encode Odorant Receptors: A Molecular Basis for Odor Recognition", *Cell*, 65, 175–187.
CALVIN W. H. and BICKERTON, D. (2000), *Lingua ex Machina*, Cambridge, The MIT Press.
CARSTAIRS-MCCARTHY, A. (1999), *The origins of complex language: an inquiry into the evolutionary beginnings of sentences, syllables and truth*, Oxford, Oxford University Press.

CAVALLI-SFORZA, L. L. (1996), *Geni, popoli e lingue*, Milano, Adelphi.
CHAMBERS, J. K. (2002), "Patterns of Variation including Change", in J. K. Chambers, P. Trudgill, and N, Schilling-Estes, *The Handbook of Language Variation and Change*, Oxford, Blackwell, 249–373.
CHO, K.W.Y. & others (1991), "Molecular nature of Spemann's organizer: the role of the *Xenopus* homeobox gene *goosecoid*", *Cell*, 67, 1111–1120.
CHOMSKY, N. (1959), "On Certain Formal Properties of Grammars", *Information and Control*, 2, 137–167.
CHOMSKY, N. (1970), "Remarks on Nominalization", in Jacobs, R. A. and Rosenbaum, P. S. (eds.) *English Transformational Grammar*. Massachusetts, Waltham, 184–221.
CHOMSKY, N. (1975), *Reflections on language*, New York, Pantheon.
CHOMSKY, N. (1981), *Lectures on government and binding*, Dordrecht, Foris.
CHOMSKY, N. (1986), *Knowledge of Language*, New York, Praeger.
CHOMSKY, N. (1996), *The Minimalist Program*, Cambridge, The MIT Press.
CHOMSKY, N. (2000), *New Horizons in the Study of Language and Mind*, Cambridge, Cambridge University Press.
CHOMSKY, N. (2002), *On Nature and Language*, ed. by A. Belletti and L. Rizzi, Cambridge, Cambridge University Press.
CHRISTIANSEN, M. & KIRBY, S. (2003), *Language evolution*, Oxford University Press.
COLLADO-VIDES, J. (1996), "Integrative Representations of the Regulation of Gene Expression", in J. Collado-Vides, B. Magasanik, and T. F. Smith (eds.), *Integrative Approahes to Molecular Biology*, Cambridge, The MIT Press, 179–203.
CRICK, F. H. C. (1970), "Central dogma of molecular biology", *Nature*, 227, 561–563.
CROFT, W. (2000), *Explaining Language Change. An Evolutionary Approach*, London, Longman.
CULICOVER, P. W. & WILKINS, W. K. (1984), *Locality in Linguistic Theory*, New York, Academic Press.
DANEŠ, F. (1967), "Order of elements and sentence intonation", *To honor Roman Jakobson*, I, The Hague, Mouton, 499–512.
DARWIN, CH. (1859), *On the Origin of Species by Means of Natural Selection*, London, John Murray.
DARWIN, CH. (1871), *The Descent of Man and Selection in Relation to Sex*, London, John Murray.
DAWKINS, R. (1976), *The selfish gene*, Oxford, Oxford University Press.
DEACON, T. W. (1997), *The Symbolic Species. The Co-evolution of Language and the Brain*, New York, Norton.
DOBZHANSKY, TH. (1937), *Genetics and the Origin of Species*, New York, Columbia University Press.
DOBZHANSKY, TH. (1975), *Genetics of the Evolutionary Process*, New York, Columbia University Press.
DOBZHANSKY, TH. (1977), "Evolution of Mankind", in Th. Dobzhansky, F. J. Ayala, G. L. Stebbins, and J. W. Valentine, *Evolution*, San Francisco, Freeman, 438–464.
DONALD, M. (1991), *Origins of the Modern Mind: Three Stages in the Evolution of Culture and Cognition*, Cambridge MA, Harvard University Press.
DONALD, M. (1997), "The mind from a historical perspective: human cognitive phylogenesis and the possibility of continuing cognitive evolution", in D. Johnson and C. Ermeling (eds.), *The Future of the Cognitive Revolution*, Oxford, Oxford University Press.

ELDREGE, N., and Gould, S. J. (1972), "Punctuated equilibria: an alternative to phyletic gradualism", in T. J. M. Schopf (ed.), *Models in Paleobiology*, San Francisco, Freeman and Cooper, 82–115.
FIRBAS, J. (1964), "On defining the Theme in Functional Sentence Analysis", *Travaux Linguistiques de Prague I*, 267–280.
FISHER, R. A. (1930), *The Genetical Theory of Natural Selection*, Oxford, Clarendon Press.
FREEMAN, W. J. (1991), "The physiology of perception", *Scientific American*, 264, 78–85.
GARDNER, R. A. & GARDNER, B. T. (1969), "Teaching sign language to a chimpanzee", *Science*, 165, 664–672.
GAZDAR, G. and others (eds.) (1983), *Order, Concord, and Constituency*, Dordrecht, Foris.
GAZDAR, G. and others (1985), *Generalized Phrase Structure Grammar*, Cambridge, Harvard University Press.
GIVON, T. (2002), *Bio-Linguistics. The Santa Barbara Lectures*, Amsterdam / Philadelphia, John Benjamins.
GOODALL, J. (1986), *The Chimpanzees of Gombe: Patterns of Behavior*, Cambridge, Harvard University Press.
GOPNIK, M. (1990), "Feature blind grammar and dysphasia", *Nature*, 344, 715.
GOULD, S. J. (1977), *Ontogeny and Philogeny*, Cambridge, Harvard University Press.
GREIMAS, A. J. (1966), *Sémantique structurale*, Paris, Larousse.
HALDANE, J. B. S. (1993), *The Causes of Evolution*, Princeton University Press.
HALLIDAY, M. A. K. & HASAN, R. (1976), *Cohesion in English*, London, Longman.
HARRIS, Z. (1951), *Structural linguistics*, Chicago, Chicago University Press.
HAUSER, M. (1996), *The Evolution of Communication*, Cambridge, The MIT Press.
HEINE, B. (1991), *Grammaticalization*, Chicago, Chicago University Press.
HEINE, B., CLAUDI, U., and HÜNNEMEYER, F. (1991), *Grammaticalization: A Conceptual Framework*, Chicago, Chicago University Press.
HELBIG, G. (1982), *Valenz – Satzglieder – Semantische Kasus – Satzmodelle*, Leipzig.
HENY, F. (ed.) (1981), *Binding and Filtering*, London, Croom Helm.
HOEY, M. (1983), *On the Surface of Discourse*, London, Allen and Unwin.
HUMBOLDT, W. von (1836), *Ueber die Verschiedenheit des menschlichen Sprachbaues*, Berlin.
JACKENDOFF, R. (1977), *X-Syntax. A Study of Phrase Structure*, Cambridge, MIT.
JACOBS, R. A. and ROSENBAUM, P. S. (1968), *English Transformational Grammar*, Waltham, Mass.
JAKOBSON, R. (1971), "Linguistics in Relation to Other Sciences", *Selected Writings II*, The Hague: Mouton, 655–696.
JENKINS, L. (2000), *Biolinguistics. Exploring the Biology of Language*, Cambridge University Press.
KAAS, J. H. (1989), "Why does the brain have so many visual areas?", *Journal of Cognitive Neuroscience*, 1.2.
KAUFFMAN, S. A. (1995), *At Home in the Universe. The Search for the Laws of Self-Organisation and Complexity*, New York, Oxford University Press.
KAUFFMAN, S. A. (2000), *Investigations*, New York, Oxford University Press.
KEGL, J., SENGHAS, A. & COPPOLA, M. (2001), "Creation through Contact: Sign Language Emergence and Sign Language Change in Nicaragua", in M. DeGraff (ed.), *Language Creation and Language Change. Creolizatin, Diachrony, and Development*, Cambridge, The MIT Press, 179–239.

KIMURA, D. (1993), *Neuromotor Mechanisms in Human Communication*, Oxford, Oxford University Press.
KNIGHT, Ch., M. STUDDERT-KENNEDY, and HURFORD, J. R., (2000), "Language: a Darwinian Adaptation?", *The Evolutionary Emergence of Language. Social Function and the Origins of Linguistic Form,* Cambridge, Cambridge University Press, 1–19.
LAMB, S. M. (1998), *Pathways of the brain. The Neurocognitive Basis of Language*, Amsterdam, John Benjamins.
LANGACKER, R (1987–1991), *Foundations of Cognitive Grammar*, Stanford University Press, 2 vols.
LENNEBERG, E. H. (1967), *Biological Foundations of Language*, New York, Wiley.
LI, Ch. N. and HOMBERT, J. M. (2002), "On the evolutionary origin of language", in M. Stamenov and V. Galese (eds.), *Mirror Neurons and the Evolution of Brain and Language*, Amsterdam, John Benjamins, 175–207.
LIEBERMANN, P. (1975), *On the Origins of Language: An Introduction to the Evolution of Human Speech*, New York, MacMillan.
LIGHTFOOT, D. (1999), *The Development of Language: Acquisition, change, and evolution*, Oxford, Blackwell.
LÓPEZ-GARCÍA, A. (1981), "Topological linguistics: liminar grammar", *Folia Linguistica*, 13–2, 267–89.
LÓPEZ-GARCÍA, A. (1990), *Introduction to Topological Linguistics*, Valencia-Minnesota, LynX.
LÓPEZ-GARCÍA, A. (1994), "Topological Linguistics and the Study of Linguistic Variation", in C. Martin (ed.), *Current Issues in Mathematical Linguistics*, Amsterdam, North-Holland, 69–77.
LÓPEZ-GARCÍA, A. (1994–1998), *Gramática del español*, Madrid, Arco, 3 vols.
LÓPEZ-GARCÍA, A. (1998), "La sintassi genetica", *Segni e Comprensione*, 32,105–128.
LÓPEZ-GARCÍA, A. (2000), *Cómo surgió el español. Introducción a la sintaxis histórica del español antiguo*, Madrid, Gredos.
LÓPEZ-GARCÍA, A. (2002), *Fundamentos genéticos del lenguaje*, Madrid, Cátedra.
LÓPEZ-GARCÍA, A. (2005), "Algunas propuestas biolingüísticas más bio- que -lingüísticas", in L. Santos (ed.), *Homenaje a F. Lázaro Carreter*, Universidad de Salamanca [forthcoming].
LORENZO, G. and LONGA, V. (2003), "Minimizing the genes for grammar. The minimalist program as a biological framework for the study of language", *Lingua*, 113, 643–657.
MARGULIS, L. (1998), *Symbiotic Planet*, New York, Basic Books.
MARR, D. (1982), *Vision*, San Francisco, Freeman.
MARTINEZ, D. E., BRIDGE, D., MASUDA, L., and CARTWRIGHT, P. (1998), "Cnidarian homeoboxes and the zootype", *Nature*, 393, 748–749.
MAYNARD SMITH, J., and SZATHMÁRY, E. (1995), *The Major Transitions in Evolution*, Oxford, Freeman.
MITHEN, S. (1996), *The Prehistory of the mind. A search for the origins of art, religion and science*, London, Thames and Hudson.
NEWMEYER, F. (1998), "On the supposed 'counterfunctionality' of Universal Grammar: some evolutionary implications", in J. R. Hurford, M. Studdert-Kennedy, and Ch. Knight (eds.), *Approaches to the Evolution of Language*, Cambridge, Cambridge University Press, 305–320.
NUNAN, D. (1993), *Introducing Discourse Analysis*, London, Penguin.
PINKER, S. and BLOOM, P. (1990), "Natural language and natural selection", *Behavioral and Brain Sciences*, 13, 707–784.

PINKER, S. (1994), *The Language Instinct. How the Mind Creates Language*, New York, William Morrow.
PREMACK, D. (1971), "Language in chimpanzee?", *Science*, 172, 808–822.
RAIBLE, W. (2003), "Linguistics and Genetics: Systematic Parallels", in I. Bréchet, Ph. Jarry et F. Létoublon (eds.), *Mécanique des signes et langage des sciences*, Grenoble, MSH-Alpes, 63–114.
RICE, M. & MAHWAH, N. J. (1996), *Towards a Genetics of Language*, New Jersey, Erlbaum.
RIDLEY, M. (1985), *The Problems of Evolution*, Oxford, Oxford University Press.
RIZZOLATTI, G., FADIGA, L., GALLESE, V., and FOGASSI, L. (1995), "Premotor cortex and the recognition of motor actions", *Cognitive Brain Research*, 3, 131–141.
RUMBAUGH, D. (1977), *Language Learning by a Chimpanzee: the Lana Project*, New York, Academic Press.
SAPIR, E. (1921), *Language. An introduction to the study of speech*, New York, Harper.
SCHEGLOFF, E. A. (1972), "Sequencing in conversational openings", in J. J. Gumperz & D. H. Hymes (eds.), *Directions in Sociolinguistics*, New York, Holt, Rinehart, and Winston, 146–180.
SCHLEICHER, A. (1873), *Die Darwinische Theorie und die Sprachwissenschaft*, Weimar [translated by A. V. Bikkers as "Darwinism tested by the science of language", in K. Koerner (ed.), *Linguistics and Evolutionary Theory*, Amsterdam, John Benjamins, 1987].
SCHRÖDINGER, E. (1944), *What is life?*, Cambridge, Cambridge University Press.
SLOBIN, D. I. (2002), "Language evolution, acquisition and diachrony: Probing the parallels", in T. Givón & B. F. Malle (eds.), *The Evolution of Language out of Pre-language*, Amsterdam, John Benjamins, 375–393.
SOLÉ, R. & B. GOODWIN (2000), *Signs of Life. How Complxity Pervades Biology*, New York, Basic Books.
SPERBER, D. & WILSON, D. (1986), *Relevance: Communication and Cognition*, Oxford, Blackwell.
STROMSWOLD, K. (2001), "The Heritability of Language: A Review and Metaanalysis of Twin, Adoption, and Linkage studies", *Language*, 77–4, 2001, 647–725.
STRYKER, M. (1990), "Activity-Dependent Reorganization of Afferents in the Developing Mammalian Visual System", in J. E, Mittenthal and A. B. Baskin (eds.), *Principles of Organization in Organisms*, Reading, Addison-Wesley.
TAMKUN, J. W. & alia (1992), "*brahma*: a regulator of *Drosophila* homeotic genes structurally related to the yeast transcriptional activator SNF2/SWI2", *Cell*, 68, 561–572.
TAYLOR, D. (1971), "Grammatical and Lexical Affinities of Creoles", in Dell Hymes (ed.), *Pidginization and Creolization of Languages*, Cambridge, Cambridge University Press, 293–296.
TESNIÈRE, L. (1959), *Eléments de syntaxe structurale*, Paris, Klincksieck.
THOM, R. (1972), *Stabilité structurelle et morphogenèse*, Paris, Interédtions.
THOM, R. (1988), *Esquisse d'une sémiophysique*, Paris, Interéditions.
TOMASELLO, M. (1999), *The Cultural Origins of Human Cognition*, Cambridge, Harvard University Press.
TRUBETZKOY, N. (1939), "Le rapport entre le déterminé, le déterminant et le défini", in *Mélanges de linguistique offerts à Charles Bally*, Génève, 75–82.
ULBAEK, I. (1998), "The origin of language and cognition", in J. R. Hurford, M. Studdert-Kennedy and Ch. Knight (eds.), *Approaches to the Evolution of Language*, Cambridge, Cambridge University Press, 30–43.

WADDINGTON, C. H. (1940), *Organizers and Genes*, Cambridge, Cambridge University Press.
WADDINGTON, C. H. (1962), *New Patterns in Genes and Development*, New York, Columbia University Press.
WIERZBICKA, A. (1996), *Semantics: Primes and Universals*, New York, Oxford University Press.
WILDGEN, W. (1999), *De la grammaire au discours. Une approche morphodynamique*, Frankfurt, Peter Lang.
WILDGEN, W. (2004), *The Evolution of Human Language. Scenarios, Principles, and Cultural Dynamics*, Berlin, John Benjamins.
WRIGHT, R. (1982), *Late Latin and Early Romance in Spain and Carolingian France*, Liverpool, Arca, F. Cairns.
ZATORRE, R. and others (ed.) (2001), *The Biological Foundations of Music*, New York, The New York Academy of Sciences, Vol. 930.
ZUBIZARRETA, M. L. (1998), *Prosody, Focus, and Word Order*, Cambridge, The MIT Press.

Index

Abler, W. 169, 170
adaptation 44, 67
agree 66
agreement 71, 72, 111, 114, 141, 164
Aiello, L.C. 12
amino acid(s) 81, 84, 96, 100, 105, 157
analogical 33, 153
analogy 33
anaphora 164
Anticodon 114
archetype 43
attractor(s) 55, 57
B-evolution 36
base(s) 97, 98, 99, 100, 141
Bickerton, D. 18, 20, 33, 63
Bloom, P. 32
Brandt, P.A. 61, 163
Calvin, W.H. 20, 33
categories 74, 101, 102, 105, 113, 164
Cavalli-Sforza, L.L. 40
Chomsky, N. 17, 21, 24, 25, 32, 52, 55, 58, 65, 68, 101, 144, 165, 170
Christiansen, M. 31, 40
code 113, 121
codon(s) 81, 84, 87, 88, 96, 99, 100, 105, 114, 123
coherence 134
complex systems 56
complexity 25, 63, 144, 145, 157
Crick, F.H.C. 95
Croft, W. 34, 35
Darwin, Ch. 15, 17, 18, 19, 21, 32, 38, 43, 44, 81, 144
Dawkins, R. 16, 30, 31, 34
Deacon, T.W. 16, 17, 18, 19, 21
dependence relations 70
discontinuities 44
DNA 91, 124, 152
Donald, M. 13, 29
Dunbar, R.I.M. 12
E-languages 64

embodied 14
emerged 29, 145
emergence 13, 18, 22, 24, 51, 144, 145, 157, 158
emergent 145
empty categories 30, 115, 116, 141
English 20, 116
enzymes 93
epigenesis 55
exaptation 18, 20, 21, 24
exon(s) 91, 124, 125, 168
figurative effect 50
focus 76, 164
fractal 145, 146, 147
gene(s) 30, 31, 34, 123
genes of grammar 24
genetic code 41, 43, 82, 83, 109, 143, 157, 172, 173
genome 90, 123, 167, 173
Gestalt 53, 54
gestaltic 61
gestaltic laws 57
Givón, T. 13
Goethe 43, 44, 64
Gould, S.J. 16
homeobox 46, 47, 48, 151, 153, 172
homology 33, 44
Humboldt, W. von 169
I-language 63, 64, 67
induction 93, 135
innate 24, 28, 171, 173
innatism 23
intron(s) 91, 124, 125, 158
Jakobson, R. 82, 83, 168, 169
Kauffman, S. 56, 57
Kirby, S. 31, 40
L-evolution 36
Lamb, S.M. 27
language faculty 28
language origin 12
lingueme 34

linguistic code 41, 82, 83, 172
Mangold, H. 149
Margulis, L. 159
memes 30, 31, 34, 40
merge 25, 52, 57, 66, 145
mimesis 29
minimal(ist) program 24, 66, 145
mirror neurons 28
Mithen, S. 13
modular 11, 147
morphemes 86
morphogenesis 148
motif(s) 151, 152, 153, 172
Motor control (theory) 29
move 30
move-α 23, 25
movement 66, 74, 109, 141
mRNA 124, 130, 168
mutation(s) 22, 24, 47, 85, 164
natural selection 15, 36, 37, 43, 67, 144, 165
neural connections 28
Nieuwkoop centre 150, 151
operon 93, 131
organizer 149, 151
origin of language 17, 29
parameters 24, 65
particulate 15
particulate principle 25, 169
phoneme 86
Pinker, S. 22, 32
polymerase 92, 125
pre-program (preprogram) 148, 155, 167
pregnance 50, 148
preRNA 124
principles 24, 65
promoter 125
protein(s) 48, 168
protolanguage(s) 18, 19, 20, 58, 61, 166
punctuated equilibrium 16, 37
purine(s) 22, 89
pyrimidine(s) 22, 89
Raible, W. 172
recursion 110, 141
replication 91, 159

repression 93, 135
Ribosome 93
Rizzolatti, G. 28
RNA 91, 152
RNA polymerase 127, 152
Saint Hilaire, Geoffroy de 44, 45, 64
salience 50, 148
satellite DNA 106, 107, 141
Schleicher, A. 33, 40
Schrödinger, E. 154
semantic roles 88
sigma factor 126
singularity 52
Spanish 20, 39, 40, 96, 116, 130
Spemann, H. 149
symbiosis 164
symbiotic 166
symbiotic unit 159
syntactic categories 96
syntactic functions 88
Tesnière, L. 112
text(s) 146, 170
theory of complexity 159
theory of form 43
theory of mind 29
Thom, R. 33, 50, 51, 52, 148
Thompson, d'Arcy W. 45, 64
Tomasello, M. 29
topological 87
topology 33
transcription 91, 124, 159
translation 92, 93, 159
transposition 141
transposon(s) 117, 118
universal grammar 21, 78, 171
variation(s) 35, 43
vision 59, 61
visual 60
visual perception 54
Wildgen, W. 33, 162
words 86
X-bar 25
X-bar structure 141
X-bar syntax 105
X-bar theory 23, 66, 145

European Semiotics: *Language, Cognition, and Culture*
Sémiotiques Européennes: *langage, cognition et culture*

Edited by / Série rédigée par
Per Aage Brandt (Aarhus), Wolfgang Wildgen (Bremen/Brême),
and/et Barend van Heusden (Groningen/Groningue)

European Semiotics originated from an initiative launched by a group of researchers in Semiotics from Denmark, Germany, Spain, France and Italy and was inspired by innovative impulses given by René Thom and his "semiophysics". The goal of the series is to provide a broad European forum for those interested in semiotic research focusing on *semiotic dynamics* and combining *cultural, linguistic and cognitive perspectives*.
This approach, which has its origins in Phenomenology, Gestalt Theory, Philosophy of Culture and Structuralism, views semiosis primarily as a cognitive process, which underlies and structures human culture. Semiotics is therefore considered to be the discipline suited *par excellence* to bridge the gap between the realms of the Cognitive Sciences and the Sciences of Culture.
The series publishes monographs, collected papers and conference proceedings of a high scholarly standard. Languages of publication are mainly English and French.

Sémiotiques européennes est le résultat d'une initiative prise par un groupe de chercheurs en sémiotique, originaires du Danemark, d'Allemagne, d'Espagne, de France et d'Italie, inspirée par l'impulsion innovatrice apportée par René Thom et sa "sémiophysique". Le but de cette collection est de fournir une tribune européenne large à tous ceux qui s'intéressent à la recherche sémiotique portant sur *les dynamiques sémiotiques*, et réunissant des *perspectives culturelles, linguistiques et cognitives*.
Cette approche, qui combine différentes sources, telle que la phénoménologie, le gestaltisme, la philosophie de la culture et le structuralisme, part du principe que la sémiosis est essentiellement un procès cognitif, qui sous-tend et structure toute culture humaine. Dans cette approche, la sémiotique est donc considérée comme la discipline par excellence capable de créer un pont entre les domaines des sciences cognitives et ceux des sciences de la culture.
Sémiotiques européennes accueille tant des monographies que des anthologies et des actes de colloques d'un haut niveau de recherche, rédigés de préférence en anglais et en français.

Volume 1	Wolfgang Wildgen: De la grammaire au discours Une approche morphodynamique xi + 351 pp. 1999. ISBN 3-906762-03-3 / US-ISBN 0-8204-4226-7
Volume 2	Lene Fogsgaard: Esquemas copulativos de SER y ESTAR Ensayo de semiolingüística 382 pp. 2000. ISBN 3-906764-22-2
Volume 3	Jean Petitot: Morphogenesis of Meaning xii + 279 pp. 2003. ISBN 3-03910-104-8 / US-ISBN 0-8204-6858-4
Volume 4	Per Aage Brandt: Spaces, Domains, and Meaning Essays in Cognitive Semiotics 271 pp. 2004. ISBN 3-03910-227-3 / US-ISBN 0-8204-7006-6
Volume 5	Marcel Bax, Barend van Heusden & Wolfgang Wildgen: Semiotic Evolution and the Dynamics of Culture xx + 318 pp. 2004. ISBN 3-03910-394-6 / US-ISBN 0-8204-9
Volume 6	Ángel López-García: The Grammar of Genes How the Genetic Code Resembles the Linguistic Code 182 pp. 2005. ISBN 3-03910-654-6 / US-ISBN 3-03910-7171-2